FUTURE U.S. WORKFORCE FOR
GEOSPATIAL INTELLIGENCE

Committee on the Future U.S. Workforce for Geospatial Intelligence

Board on Earth Sciences and Resources
Division on Earth and Life Studies

Board on Higher Education and Workforce
Policy and Global Affairs

NATIONAL RESEARCH COUNCIL
OF THE NATIONAL ACADEMIES

THE NATIONAL ACADEMIES PRESS
Washington, D.C.
www.nap.edu

THE NATIONAL ACADEMIES PRESS • 500 Fifth Street, NW • Washington, DC 20001

NOTICE: The project that is the subject of this report was approved by the Governing Board of the National Research Council, whose members are drawn from the councils of the National Academy of Sciences, the National Academy of Engineering, and the Institute of Medicine. The members of the committee responsible for the report were chosen for their special competences and with regard for appropriate balance.

This study was supported by Contract No. HM1582-09-C-0014 between the National Academy of Sciences and the National Geospatial-Intelligence Agency. The views, opinions, and findings contained in this report are those of the author(s) and should not be construed as an official Department of Defense position, policy, or decision unless so designated by other official documentation.

International Standard Book Number-13: 978-0-309-26864-6
International Standard Book Number-10: 0-309-26864-8

Cover illustrations: Photo of a battlefield use of sensor input to a mobile display device, courtesy of DARPA (upper left); satellite image of Afghanistan, courtesy of USGS (upper right); *Map of Science*, courtesy of Richard Klavans, used with permission from SciTech (background map).

Additional copies of this report are available for sale from the National Academies Press, 500 Fifth Street, NW, Keck 360, Washington, DC 20001; (800) 624-6242 or (202) 334-3313; http://www.nap.edu/.

THE NATIONAL ACADEMIES
Advisers to the Nation on Science, Engineering, and Medicine

The **National Academy of Sciences** is a private, nonprofit, self-perpetuating society of distinguished scholars engaged in scientific and engineering research, dedicated to the furtherance of science and technology and to their use for the general welfare. Upon the authority of the charter granted to it by the Congress in 1863, the Academy has a mandate that requires it to advise the federal government on scientific and technical matters. Dr. Ralph J. Cicerone is president of the National Academy of Sciences.

The **National Academy of Engineering** was established in 1964, under the charter of the National Academy of Sciences, as a parallel organization of outstanding engineers. It is autonomous in its administration and in the selection of its members, sharing with the National Academy of Sciences the responsibility for advising the federal government. The National Academy of Engineering also sponsors engineering programs aimed at meeting national needs, encourages education and research, and recognizes the superior achievements of engineers. Dr. Charles M. Vest is president of the National Academy of Engineering.

The **Institute of Medicine** was established in 1970 by the National Academy of Sciences to secure the services of eminent members of appropriate professions in the examination of policy matters pertaining to the health of the public. The Institute acts under the responsibility given to the National Academy of Sciences by its congressional charter to be an adviser to the federal government and, upon its own initiative, to identify issues of medical care, research, and education. Dr. Harvey V. Fineberg is president of the Institute of Medicine.

The **National Research Council** was organized by the National Academy of Sciences in 1916 to associate the broad community of science and technology with the Academy's purposes of furthering knowledge and advising the federal government. Functioning in accordance with general policies determined by the Academy, the Council has become the principal operating agency of both the National Academy of Sciences and the National Academy of Engineering in providing services to the government, the public, and the scientific and engineering communities. The Council is administered jointly by both Academies and the Institute of Medicine. Dr. Ralph J. Cicerone and Dr. Charles M. Vest are chair and vice chair, respectively, of the National Research Council.

www.national-academies.org

Acknowledgments

This report has been reviewed in draft form by individuals chosen for their diverse perspectives and technical expertise, in accordance with procedures approved by the NRC's Report Review Committee. The purpose of this independent review is to provide candid and critical comments that will assist the institution in making its published report as sound as possible and to ensure that the report meets institutional standards for objectivity, evidence, and responsiveness to the study charge. The review comments and draft manuscript remain confidential to protect the integrity of the deliberative process. We wish to thank the following individuals for their participation in the review of this report:

Burt S. Barnow, The George Washington University, Washington, D.C.

Katy Börner, Indiana University, Bloomington

Gaetano Borriello, University of Washington, Seattle

Cynthia A. Brewer, Pennsylvania State University, University Park

Joseph Galaskiewicz, University of Arizona, Tucson

Ayman F. Habib, University of Calgary, Canada

Michael J. Jackson, University of Nottingham, United Kingdom

Annette J. Krygiel, Independent Consultant, Great Falls, Virginia

James Llinas, State University of New York at Buffalo

Marguerite Madden, University of Georgia, Athens

Leif E. Peterson, Advanced HR Concepts & Solutions, LLC, Beavercreek, Ohio

Chris Rizos, University of New South Wales, Australia

Scott A. Sandgathe, University of Washington, Seattle

Stephen M. Stigler, University of Chicago, Illinois

Although the reviewers listed above have provided many constructive comments and suggestions, they were not asked to endorse the conclusions or recommendations nor did they see the final draft of the report before its release. The review of this report was overseen by David R. Rain, George Washington University, and Stephen R. Berry, University of Chicago. Appointed by the National Research Council, they were responsible for making certain that an independent examination of this report was carried out in accordance with institutional procedures and that all review comments were carefully considered. Responsibility for the final content of this report rests entirely with the authoring committee and the institution.

The committee would like to thank the following individuals who gave briefings or provided data, figures, or other input: Max Baber, Richard Berg, Adam Chill, Denise Clayton Delahanty, David DiBiase, Michael Gibbons, Leila Gonzales, George Hepner, Susan Kalweit, Nirmala Kannankutty, Keith Krut, Keith Masback, Patricia Muth, Wendy Nelson, Mark Pahls, Jürgen Pfeffer, James Plasker, Mark Regets, Jack Sanders, H. Greg Smith, Marc Spraragen, George Stamus, Ed Waltz, and Michael Wolf.

Contents

Summary

We live in a changing world with multiple and evolving threats to national security, including terrorism, asymmetrical warfare (conflicts between agents with different military powers or tactics), and social unrest. Visually depicting and assessing these threats using imagery and other geographically referenced information is the mission of the National Geospatial-Intelligence Agency (NGA). As the nature of the threat evolves, so do the tools, knowledge, and skills needed to respond. Technological advances are moving geospatial tools and near-real-time information products into the hands of warfighters, emergency responders, and other users. New geospatial themes and interdisciplinary approaches to problem solving that could potentially improve geospatial intelligence (GEOINT) are emerging in university curricula. In addition, a new generation of students accustomed to working in flexible, socially connected, and highly integrated technological environments is bringing new capabilities into the workplace.

The challenge for NGA is to maintain a workforce that can deal with evolving threats to national security, ongoing scientific and technological advances, and changing skills and expectations of workers. The agency's success depends in part on the availability of experts with suitable knowledge and skills. At the request of H. Greg Smith, NGA chief scientist, the National Research Council (NRC) established a committee to assess the supply of expertise in geospatial intelligence fields, identify gaps in expertise relative to NGA's needs, and suggest ways to ensure an adequate supply of geospatial intelligence expertise over the next 20 years (see Box S.1).

This report analyzes the geospatial intelligence workforce in 10 areas defined in *New Research Directions for the National-Geospatial Intelligence Agency: Workshop Report* (NRC, 2010a), including 5 core areas (geodesy and geophysics, photogrammetry, remote sensing, cartographic science, geographic information systems [GIS] and geospatial analysis) and 5 emerging areas (GEOINT fusion, crowdsourcing, human geography, visual analytics, and forecasting). The availability of expertise in these areas was assessed using education and labor statistics collected from government sources. Gaps in expertise relative to NGA's needs were identified by comparing the statistics to information on NGA's current scientist and analyst positions and published assessments of demand for geospatial occupations. Ideas for building the necessary knowledge and skills were chosen based on a review of training programs in universities, professional societies, government agencies, and private companies.

GEOSPATIAL INTELLIGENCE FIELDS

NGA scientists and analysts use imagery and geospatial data to depict features and activities on, above, or below the surface of the Earth to help users visualize what is happening and where. The current production and analysis of geospatial intelligence relies primarily on the techniques of the five core areas:

- *Geodesy and geophysics*—Geodesy is the science of mathematically determining the size, shape, and orientation of the Earth and the nature of its gravity field in four dimensions. It includes the development

---

BOX S.1
Committee Charge

An ad hoc committee will examine the need for geospatial intelligence expertise in the United States compared with the production of experts in the relevant disciplines, and discuss possible ways to ensure adequate availability of the needed expertise. In its report the committee will

1. Examine the current availability of U.S. experts in geospatial intelligence disciplines and approaches and the anticipated U.S. availability of this expertise for the next 20 years. The disciplines and approaches to be considered include NGA's five core areas and promising research areas identified in the May 2010 NRC workshop.
2. Identify any gaps in the current or future availability of this expertise relative to NGA's need.
3. Describe U.S. academic, government laboratory, industry, and professional society training programs for geospatial intelligence disciplines and analytical skills.
4. Suggest ways to build the necessary knowledge and skills to ensure an adequate U.S. supply of geospatial intelligence experts for the next 20 years, including NGA intramural training programs or NGA support for training programs in other venues.

The report will not include recommendations on policy issues such as funding, the creation of new programs or initiatives, or government organization.

---

of highly precise positioning techniques and monitoring of dynamic Earth phenomena. Geophysics is the physics of the Earth and its environment in space, including the study of geodesy, geomagnetism and paleomagnetism, seismology, hydrology, space physics and aeronomy, tectonophysics, and atmospheric science.

- *Photogrammetry*—the art, science, and technology of extracting reliable and accurate information about objects, phenomena, and environments from the processing of acquired imagery and other sensed data, both passively and actively, within a wide range of the electromagnetic energy spectrum.
- *Remote sensing*—the science of measuring some property of an object or phenomenon by a sensor that is not in physical contact with the object or phenomenon under study.
- *Cartographic science*—the discipline dealing with the conception, production, dissemination, and study of maps as both tangible and digital objects, and with their use and analysis.

- *Geographic Information Systems and geospatial analysis*—GIS refers to any system that captures, stores, analyzes, manages, and visualizes data that are linked to location. Geospatial analysis is the process of applying analytical techniques to geographically referenced data sets to extract or generate new geographical information or insight.

Recently, five research areas have emerged in academia that could improve geospatial intelligence by adding new types of information and analysis methods as well as new capabilities to help anticipate future threats:

- *GEOINT fusion*—the aggregation, integration, and conflation of geospatial data across time and space with the goal of removing the effects of data measurement systems and facilitating spatial analysis and synthesis across information sources.
- *Crowdsourcing*—a process in which individuals gather and analyze information and complete tasks over the Internet, often using mobile devices such as cellular phones. Individuals with these devices form interactive, scalable sensor networks that enable professionals and the public to gather, analyze, share, and visualize local knowledge and observations and to collaborate on the design, assessment, and testing of devices and results.
- *Human geography*—the science of understanding, representing, and forecasting activities of individuals, groups, organizations, and the social networks to which they belong within a geotemporal context. It includes the creation of operational technologies based on societal, cultural, religious, tribal, historical, and linguistic knowledge; local economy and infrastructure; and knowledge about evolving threats within that geotemporal window.
- *Visual analytics*—the science of analytic reasoning, facilitated by interactive visual interfaces. The techniques are used to synthesize information and derive insight from massive, dynamic, ambiguous, and often conflicting data.
- *Forecasting*—an operational research technique used to anticipate outcomes, trends, or expected future behavior of a system using statistics and modeling. A forecast is used as a basis for planning and decision making and is stated in less certain terms than a prediction.

EVOLUTION OF THE CORE AND EMERGING AREAS

Education and training in the core and emerging areas is provided primarily by universities and colleges, so the evolution of these areas as academic endeavors directly influences the supply of graduates with NGA-relevant skills. Disciplinary change has significantly modified the content and educational profile of NGA's core areas over the past several decades. For example, GIS has been transformed from software systems developed by a few commercial vendors to a wide range of web services supported by open standards. The focus of geospatial analysis has shifted from supporting GIS applications to using space-time analytic measures and large amounts of data to study the dynamics of human and physical systems. Advances in sensors and image processing are yielding increasingly detailed remote sensing imagery, and sensors are starting to be linked into sensor webs, which offer new ways to monitor and explore environments remotely. More and better sensors and improved processing capabilities are also producing more detailed images of the Earth's interior and its magnetic and gravitational fields. The resulting changes in curricula have generally taken place within the traditional university departments for remote sensing, geophysics, and GIS and geospatial analysis.

In contrast, disciplinary change in the other core areas has led to name changes, overlaps in content or methods, and/or moves to different departments. Digital imagery and automated processing have brought the methods of digital photogrammetry close to those of remote sensing. The digital transition has profoundly affected cartography by providing online methods (e.g., interactive maps) and new graphical techniques (e.g., geovisualizations) to illustrate and communicate spatial information beyond the paper map. In response, university curricula have shifted from cartography to geographic information science, a broader field encompassing the science and technology of geographic information. Traditional cartographic training in map production and the principles of graphic display have been replaced by training to analyze spatial patterns and to represent them effectively on maps and charts, often using GIS. This shift has narrowed the differences among cartography, GIS, and geospatial analysis.

Geodesy and photogrammetry were used extensively by military and intelligence agencies in the 1960s, 1970s, and 1980s. Automation and the increased use of other methods (e.g., remote sensing, geospatial analysis) led to substantial reductions both in the number of photogrammetry and geodesy specialists in military and intelligence occupations and in the academic programs that produced them. Most important is the decline in photogrammetry, which is in danger of disappearing as a specialized course of study in universities. Geodesy, which underpins a wide range of civil applications (e.g., surveying, navigation, environmental monitoring), continues to be taught at several universities, although degrees are offered mainly at the master's and doctorate levels. At the undergraduate level, geodesy and photogrammetry have largely been incorporated into geomatics programs, which cover the science, engineering, and art of collecting and managing geographically referenced information.

By their nature, the emerging areas are still developing as areas of research and training, and the academic infrastructure (e.g., professional societies, journals) to support their development is in its infancy. Only a handful of universities offer research programs in emerging areas and even fewer offer degree programs. Most of the programs are interdisciplinary, and student training is provided largely through individual courses often scattered among different university departments.

SUPPLY OF EXPERTISE IN GEOSPATIAL INTELLIGENCE FIELDS

The first task of the committee was to estimate the supply of experts in the core and emerging areas now and over the next 20 years. NGA draws on two sources of experts for its scientist and analyst positions: (1) new graduates in relevant fields of study, and (2) individuals working in occupations that require similar knowledge and/or skills. The committee obtained statistics on these sources from the Department of Education, which tracks the number of degrees conferred in more than 1,000 fields of study, and by the Bureau of Labor Statistics, which tracks the number of jobs in more than 800 occupations. Unfortunately, the statistics are not ideal for addressing the task because the core and emerging areas are either embedded within broader fields of study and occupations or they span several fields of study or occupations. Only one field of study (cartography) and no occupations directly match the core and emerging areas. Consequently, the committee made two estimates of the number of experts:

1. An "upper-bound" estimate encompassing new graduates in all potentially relevant fields of study and workers in all potentially relevant occupations.[1] These individuals likely have some knowledge and skills relevant to a core or emerging area, and could potentially be trained for a science or analyst position at NGA.

2. A much lower estimate of the number of new graduates and workers with education or experience in a core or emerging area. These individuals may possess the desired mix of knowledge and skills without the need for substantial on-the-job training.

Supply of New Graduates and Workers with Some Relevant Skills

For the "upper-bound" estimate, the committee chose 109 fields of study and 36 occupations that are highly relevant to the core and emerging areas, and then summed the number of graduates and workers who are U.S. citizens and permanent residents. NGA's requirement for U.S. citizenship reduces the pool of new graduates by 7 percent, with the largest reductions at the doctorate level, and the pool of experienced workers by 12 percent, with the largest reductions in physical science and computer occupations. The statistics show that U.S. citizens and permanent residents received more than 200,000 degrees in relevant fields of study in 2009, and that U.S. citizens held more than 2.4 million jobs in relevant occupations in 2010, the latest years for which statistics were available when this report was written.

The future supply of geospatial intelligence experts depends primarily on the number of people graduating with degrees in relevant fields of study. To estimate an "upper bound" on the number of graduates over the next 20 years, the committee extrapolated 10-year trends in the number of graduates in the 109 fields of study. The uncertainty in the estimate was characterized by extrapolating the number of new graduates under a high-growth scenario (50 percent higher than the

growth rate observed for 2000–2009) and a low-growth scenario (50 percent lower than the observed growth rate). The results suggest that between 312,000 and 648,000 degrees in relevant fields of study will be conferred to U.S. citizens and permanent residents in 2030.

Supply of New Graduates and Workers with Education or Training in a Core or Emerging Area

To estimate the number of new graduates with education in a core or emerging area, the committee used expert judgment to weigh the education statistics against other factors, including the number of universities offering programs in a core or emerging area, the instructional programs that produce the bulk of necessary skills, and the number of members in key professional societies. Factoring in this information yields a current number of graduates on the order of tens for photogrammetry; tens to hundreds for GEOINT fusion, crowdsourcing, human geography, and visual analytics; hundreds for geodesy, geophysics, and cartographic science; hundreds to thousands for remote sensing and forecasting; and thousands for GIS and geospatial analysis. Although accurate projections of these qualitative estimates cannot be made, past trends suggest that the number of graduates will rise over the next 20 years in all areas except photogrammetry and cartography.

Estimates of the number of workers experienced in a core or emerging area cannot be made from the broad occupation categories tracked by the Bureau of Labor Statistics. The numbers are likely low in the emerging areas because the supply of graduates has been low. For the core areas, a "lower bound" was estimated by summing the number of jobs in the four most closely related occupations: cartographers and photogrammetrists; surveying and mapping technicians; geographers; and geoscientists, except hydrologists and geographers. In 2010, there were nearly 100,000 jobs in these occupations, approximately 4 percent of the "upper-bound" estimate.

Answer to Task 1

The education and labor analysis suggests that the current number of U.S. citizens and permanent residents with education in a core or emerging area is likely on the order of tens for photogrammetry; tens to

---

[1] Estimates were based on the 2000 version of the Department of Education's Classification of Instructional Programs and the 2010 version of the Bureau of Labor Statistics' occupational codes.

hundreds for GEOINT fusion, crowdsourcing, human geography, and visual analytics; hundreds for geodesy, geophysics, and cartographic science; hundreds to thousands for remote sensing and forecasting; and thousands for GIS and geospatial analysis. In addition, U.S. citizens currently hold more than 100,000 jobs in occupations closely related to the core areas. If substantial on-the-job training is an option for NGA, the current labor pool increases to 200,000 new graduates and 2.4 million experienced workers. If 10-year growth trends in the "upper-bound" estimate continue, the number of new graduates could reach 312,000–649,000 by 2030.

GAPS IN EXPERTISE RELATIVE TO NGA'S NEEDS

The second task of the committee was to identify gaps in the availability of geospatial intelligence expertise relative to NGA's needs. The expertise available to NGA depends not only on the supply of new graduates and experienced workers (discussed above) but also on the demand for knowledge and skills by NGA and other organizations. Demand for expertise by other organizations was estimated from published studies on the geospatial industry. NGA's current needs were characterized from the number of employees in various scientist and analyst occupations, the degrees and coursework specified in NGA occupation descriptions, and the types of training offered to new employees through the NGA College. Strategic information, such as current problems finding expertise and future hiring priorities, were not available from NGA, so the committee made two assumptions: (1) that the NGA College curriculum reflects not only what topics are currently important to NGA, but also what knowledge and skills are hard to find in applicants; and (2) that NGA currently needs expertise in the five core areas and that the five emerging areas would become increasingly important in the future. Based on this information and the assumptions, the committee identified gaps in domain knowledge and skills and where to find them.

Domain Knowledge

The committee identified gaps in domain knowledge relative to NGA's needs by comparing the number of experts (new graduates with education in a core or emerging area and experienced workers in closely related occupations) with the number of scientists and analysts hired by NGA (historically several hundred per year) and their areas of expertise. The largest fractions of NGA scientists and analysts work on imagery analysis (40 percent), geospatial analysis (19 percent), and cartography (10 percent).

The comparison shows that the number of graduates and experienced workers exceeds the small number of NGA positions in all core areas. Expertise in geophysics and geospatial analysis is likely sufficient for NGA's current and future needs. NGA hires only a small fraction of the available experts and offers little or no training in these areas to employees through the NGA College. There appear to be enough cartographers, photogrammetrists, and geodesists for NGA's current needs. The number of professionals working in these areas is substantially higher than the number of NGA positions, and only minimal training is offered at the NGA College. However, future shortages in cartography, photogrammetry, and geodesy seem likely because the number of graduates is too small (tens to hundreds) to give NGA choices or means of meeting sudden demand. Moreover, cartography and photogrammetry programs are shrinking. Some shortages may be imminent, given that industry is already having trouble filling cartography positions and that federal agencies are concerned about a growing deficit of highly skilled geodesists. It is possible that GIS and remote sensing recruits are already hard to find, given the extensive training in these fields provided by the NGA College. Although the supply in both fields exceeds NGA's needs, competition for GIS applications analysts is strong.

NGA has no positions in emerging areas, so any gaps in expertise will occur in the future. Emerging areas are likely to become increasingly important to NGA, in part because they are based on interdisciplinary approaches, which are needed to tackle big data and complex intelligence problems, such as those that concern coupled human-environmental systems. Such interdisciplinary approaches are also useful for many other applications, so competition, coupled with a small supply (tens to hundreds in most emerging areas), could lead to shortages in the future availability of expertise in the emerging areas.

Skills

NGA occupation descriptions specify both core competencies for all science and analyst positions and those skills required for each type of position. The core competencies stress interpersonal skills, communication, and creative thinking and adaptability, whereas the position-related skills stress working with customers and gathering, analyzing, and disseminating information. The NGA College offers several courses in interpersonal skills, written and oral communication, and critical thinking, suggesting that these core competencies are currently in short supply.

In the foreseeable future, new questions, as well as the data sets and tools needed to answer them, will continually arise. Dealing with these evolving questions and approaches requires a flexible workforce that is capable of thinking in breadth, rather than depth, through interdisciplinary training and teamwork. The ideal skill set will include spatial thinking, scientific and computer literacy, mathematics and statistics, languages and world culture, and professional ethics. Some of these skills (statistics, ethics, cultural analysis, and scientific methods) are required for particular NGA positions. Although many university programs teach some of these skills, graduates with the ideal skill set are scarce. In particular, math and computer skills remain a gap in many natural and social science programs, and spatial skills remain a gap in many computer science and engineering programs. These gaps are likely to persist until more interdisciplinary programs develop.

Recruiting

Individuals with knowledge and skills in the core and emerging areas are available, but NGA may not be looking for them in all the right places. NGA focuses recruiting on academic institutions that are near major NGA facilities or that have a large population of underrepresented groups. Only about one-third of these institutions, typically the large state universities, have strong programs in core or emerging areas, although many likely help meet other agency goals, such as increasing diversity. Extending recruiting to some of the example university programs identified in this report would help NGA find the geospatial intelligence expertise it needs.

Answer to Task 2

The committee's analysis revealed both current and future gaps in knowledge and skills relative to NGA's needs. Although the supply of experts is larger than NGA demand in all core and emerging areas, qualified GIS and remote sensing experts may already be hard to find. Long before 2030, competition and a small number of graduates will likely result in shortages in cartography, photogrammetry, geodesy, and all emerging areas. In NGA's future workforce, which is likely to be more interdisciplinary and focused on emerging areas, the ideal skill set will include spatial thinking, scientific and computer literacy, mathematics and statistics, languages and world culture, and professional ethics. Although NGA is currently finding employees with skills in statistics, ethics, cultural analysis, and scientific methods, graduates with the ideal skill set will remain scarce until interdisciplinary and emerging areas develop. NGA could improve its chances of finding the necessary knowledge and skills by extending recruiting to the example university programs identified in this report.

CURRENT TRAINING PROGRAMS

Answer to Task 3

The third task of the committee was to describe training programs relevant to geospatial intelligence that are offered by a variety of organizations. The committee chose example programs that have a long record of accomplishment, a critical mass of high-caliber instructors, a substantial number of students, and/or that provide an opportunity to solve problems in a real-world context. Universities provide the foundation knowledge and skills needed by NGA scientists and analysts. Degree programs offer comprehensive coursework in a field of study (e.g., University of Colorado's Department of Geography), as well as important supporting classes, such as statistics and mathematics. Some university programs teach the ability to think and work across disciplinary boundaries (e.g., Carnegie Mellon University's Computational and Organization Science program), to combine scientific knowledge with practical workforce skills (e.g., North Carolina State University's professional science master's in geospatial information science and technology), or to apply scientific knowledge to solve real-world problems (e.g., George Mason University's master's in geographic and cartographic sciences), sometimes in the context of national security and defense (e.g., military colleges). Other organizations offer short-term, immersive training, which is particularly useful for updating or augmenting employee skills. Courses offered by government agencies are usually targeted at agency operational needs (e.g., National Weather Service's Warning Decision Training Branch). Short courses and conference workshops offered by professional societies and other nongovernmental organizations provide focused training and sometimes certificates on specific geospatial topics (e.g., Institute of Navigation's short courses in positioning, navigation, and timing). Private companies commonly provide training for using the software (e.g., Environmental Systems Research Institute's [ESRI's] GIS software) and hardware (e.g., Gloal Positioning System receivers, photogrammetric workstations) they have developed.

WAYS TO BUILD KNOWLEDGE AND SKILLS IN THE FUTURE

The fourth task of the committee was to suggest ways to build the necessary knowledge and skills to ensure an adequate U.S. supply of geospatial intelligence experts over the next 20 years. Few of the training programs mentioned above were designed specifically for NGA's employment needs and, thus, do not offer all of the knowledge and skills needed by the agency. However, a variety of mechanisms are available for NGA to build the specialized expertise it needs in the future, including strengthening existing training programs, building core and emerging areas, and enhancing recruiting. A menu of options, of varying scope and complexity, that NGA is not currently utilizing is described below.

Strengthening Training

NGA uses existing training programs to obtain knowledge and skills, but some of these programs could be strengthened to better meet the agency's needs. For example, in addition to sending employees to short courses at professional society conferences, NGA could encourage university professors to develop short courses in emerging areas or other subjects of interest to NGA. Setting up short courses, workshops, and seminars is relatively simple, requiring only credentialed instructors and an event organizer.

NGA seeks university training for new employees and also sends some employees to universities for advanced training in core areas through the Vector Study Program. The program allows NGA employees to attend school for three semesters (undergraduate study) or six semesters (graduate study) while receiving full salary and benefits. However, university training through the Vector Study Program is being replaced by less in-depth training at the NGA College. Increasing the number of employees who participate in the Vector Study Program would enhance employee skills in core areas, and extending the program to emerging areas would bring new skills to the agency. Allowing distance learning or shorter or longer periods of study would make the program more flexible to both NGA and its employees.

Finally, the NGA College offers approximately 170 courses to its employees and other government workers and contractors. Courses are taught by government employees and contractors. External reviews by independent experts, which are common in university departments, would help administrators ensure that the curriculum remains relevant and up to date and that the teaching staff are of the highest caliber.

Building Core and Emerging Areas

NGA provides grants to academic institutions and consortia to support research and education in geospatial intelligence fields. Grant programs could also be used to support core and emerging areas by establishing research centers and partnerships and by helping to develop curricula and academic support infrastructure. Centers provide a means to gather experts from different fields and/or different organizations to develop new research areas. They can take several forms, depending on the goals and partners in the collaboration. Government research centers attached to a university (University Affiliated Research Centers [UARCs]) are established to help an agency maintain core scientific and technologic capabilities over a long period. Research centers and partnerships may also be established between private companies and universities and/or government agencies to support technological innovation. Centers of excellence can be housed in a university, federal agency, or private company and can focus on any topic that requires a team approach or shared facilities. They are commonly established to carry out collaborative research, create tools and data sets, and build a cohort of trained individuals in new subject areas. In virtual centers, members work together from their own institutions using conferencing and the Internet. They are easy to set up and are often established to facilitate work on short-term projects or new research areas.

By supporting university research, NGA indirectly influences the development of fields of interest. NGA could speed the development of emerging areas by sponsoring university efforts to establish core curricula and academic support infrastructure (i.e., journals, professional societies). Core curricula are particularly important in emerging areas because each program has a unique set of collaborating departments and approaches for dealing with the topics, so graduates from different programs commonly have different knowledge and skills. The academic support infrastructure for the emerging areas could be nurtured through actions such as funding a university scientist to compile and edit a special issue on an emerging topic in a leading journal or organizing sessions on emerging themes at key conferences.

Enhancing Recruiting

NGA offers scholarships and internships to support students interested in a career in geospatial intelligence. Other ways to reach potential applicants include organizing sessions at professional society conferences to raise awareness of NGA and its technical work, and establishing a social media site with links to job listings, recruiting events, and related information to make it easy to find information about NGA careers. NGA could seek candidates with the right combination of spatial reasoning skills by engaging students in interesting problem-solving exercises (e.g., analyzing an intelligence problem) at recruiting events. In addition, career aptitude tests, administered by NGA or by various testing services, could be used to find individuals

with abilities in spatial thinking, geography, or image interpretation.

Answer to Task 4

The actions described above to answer Task 4 show that a variety of mechanisms can be used to ensure the future availability of geospatial intelligence expertise. Some of the mechanisms would build expertise in the long term (e.g., UARCs, research partnerships with industry, curriculum development, academic support infrastructure), while others could provide more immediate gains (e.g., Vector Study Program expansion, virtual centers, professional society workshops and short courses, recruitment efforts). Most mechanisms would be relatively inexpensive to implement (e.g., virtual centers, curriculum development, recruiting efforts), while some could require substantial investment, depending on size and scope (e.g., UARCs, Vector Study Program expansion, centers of excellence). The need is greatest for the emerging areas, which currently produce few graduates and lack the academic infrastructure to develop quickly, but these mechanisms could also be used to build other areas of interest to NGA. Getting involved with education and training programs would also provide opportunities for NGA to influence the development of fields it relies on to carry out its mission.

The bottom line is that, despite its need for highly specialized knowledge and skills, NGA has the comparative luxury of being a small employer in the burgeoning geospatial enterprise. NGA is probably finding sufficient experts in all core areas, with the possible exception of GIS and remote sensing. However, shortages (too few experts to give NGA choices or means of meeting sudden demand) in photogrammetry, cartography, and geodesy are likely in the short term, followed by possible shortages in emerging areas in the longer term. While low numbers of experts are of concern to NGA, many mechanisms are available to build the knowledge and skills that NGA will require, such as strengthening existing training programs, building core and emerging areas, and enhancing recruiting. With attention to these areas, NGA has the ability both to meet its workforce needs and to be adaptive to a changing mission during the next 20 years, and potentially well beyond.

1

Introduction

The National Geospatial-Intelligence Agency (NGA) is responsible for providing timely, relevant, and accurate imagery, geospatial information, and products—collectively known as geospatial intelligence—to support national security. The threats to national security continually evolve, as do the tools and skill sets needed to respond. As a result, NGA faces the challenge of maintaining a workforce that can deal with changes in the location of conflicts, the nature of warfare (Münkler, 2003), and the management of asymmetrical threats (conflicts between agents with different military powers or tactics; Geiss, 2006), as well as ongoing scientific and technological advances, competition for geospatial expertise by other organizations, and the changing expectations of workers.

NGA scientists and analysts use imagery and geospatial information to describe, assess, and visually depict physical features and geographically referenced activities on the Earth. To carry out this work, NGA has historically hired individuals in five core areas: geodesy and geophysics, photogrammetry, remote sensing, cartographic science, and Geographic Information Systems (GIS) and geospatial analysis. These five fields have also been at the core of the commercial geospatial sector in the United States over the past decade (e.g., Google Earth, mobile location-based services). However, university programs, which provide foundation geospatial knowledge and skills, are constantly changing, as are the skill sets of graduates.

At the same time, recent technological shifts—including open-source data exploitation, crowdsourcing, distributed computing, and hand-held mobile devices—are moving more geospatial intelligence tools and products into the hands of the warfighter and, in doing so, are changing the nature of the work done at the NGA. These technological advances are also generating new geospatially oriented businesses (e.g., FourSquare, Groundspeak) and influencing academic programs. New geospatial themes are emerging in university curricula—including geospatial intelligence fusion, crowdsourcing, human geography, visual analytics, and forecasting—that could potentially improve the quality and timeliness of geospatial intelligence (NRC, 2010a). Many of these new fields take advantage of the software and networking skills of students in the millennium generation, who are technologically savvy compared to their peers a few decades ago. Moreover, new programs in universities are beginning to yield students with knowledge across multiple fields, potentially bringing new approaches to geospatial intelligence. Universities increasingly offer interdisciplinary degree programs, such as a computer science major with a GIS emphasis. The use of spatial reasoning and visualization for problem solving is now a feature of many academic programs beyond the traditional field of geography.

Although the overall supply of geospatial experts is growing, so too is the demand for these experts from other agencies and the private sector (e.g., Gewin, 2004; DiBiase et al., 2006; Solem et al., 2008). Consequently, NGA is competing with other organizations for specialists with geospatial skills. At the request of H. Greg Smith, NGA Chief Scientist, the National Research Council established an expert committee to examine the supply of experts in geospatial intelligence disciplines and to suggest ways for NGA to obtain the scientific knowledge and analytical skills it needs over the next 20 years. The specific charge to the committee is given in Box 1.1.

COMMITTEE APPROACH

This report is the second of two requested by NGA. The first report, *New Research Directions for the National Geospatial-Intelligence Agency: Workshop Report* (NRC, 2010a), summarized workshop discussions of new research directions for geospatial intelligence. The workshop considered 10 subject areas, including NGA's five core areas and five crosscutting themes that are likely to become increasingly important to NGA over the next 15 years. Definitions of these areas, slightly refined from those given in NRC (2010a), are given in Box 1.2. This report builds from the workshop results, analyzing workforce trends and education and training programs in the 10 core and emerging areas.

The committee began its analysis by characterizing the 10 core and emerging areas, including their evolution, the scope of university programs offering classes and/or degrees, and the body of knowledge and skills that are generally taught. Information for this overview was drawn from professional societies, university websites, and the committee members' own knowledge and experience. Next, the committee assessed the availability of experts in the core and emerging areas over the next 20 years (Task 1). The committee considered two sources of potential employees for NGA: (1) new graduates entering the workforce and (2) individuals currently employed in occupations that require similar knowledge and/or skills. Statistics on new graduates were obtained from the Department of Education, which tracks the number of degrees conferred by level and field of study and by citizenship. Employment statistics for more than 800 occupations were obtained from the Department of Labor's Bureau of Labor Statistics, and citizenship of employed individuals was determined from Census data. Based on the education and skill requirements laid out in NGA occupation descriptions and the committee's evaluation of the core and emerging areas, 164 instructional programs (Appendix C) and 36 occupations (Appendix D) were deemed relevant to NGA. Although a few professional societies collect degree and employment information for some of the subject areas (e.g., geophysics, photogrammetry, remote sensing), the data are less comprehensive and consistent than the government statistics and were not analyzed in this report.

For Task 2, the committee was asked to identify gaps in the current or future availability of geospatial intelligence expertise relative to NGA's needs. NGA's current needs were characterized using information provided by the agency or posted on its website (Box 1.3). The NGA job listings and position descriptions provide a measure of the knowledge and skills the agency is currently seeking, and the universities where NGA recruits provide an indication of where the agency is looking for this knowledge and skills. Based on discussions with NGA managers, the committee focused on science and analysis positions (Box 1.4), not on management or support positions (e.g., administrative assistants, database administrators). Future needs were estimated from the age distribution of agency scientists and analysts and the assumption that hiring

---

> ## BOX 1.1
> ## Committee Charge
>
> An ad hoc committee will examine the need for geospatial intelligence expertise in the United States compared with the production of experts in the relevant disciplines, and discuss possible ways to ensure adequate availability of the needed expertise. In its report the committee will
>
> 1. Examine the current availability of U.S. experts in geospatial intelligence disciplines and approaches and the anticipated U.S. availability of this expertise for the next 20 years. The disciplines and approaches to be considered include NGA's five core areas and promising research areas identified in the May 2010 NRC workshop [see Box 1.2].
> 2. Identify any gaps in the current or future availability of this expertise relative to NGA's need.
> 3. Describe U.S. academic, government laboratory, industry, and professional society training programs for geospatial intelligence disciplines and analytical skills.
> 4. Suggest ways to build the necessary knowledge and skills to ensure an adequate U.S. supply of geospatial intelligence experts for the next 20 years, including NGA intramural training programs or NGA support for training programs in other venues.
>
> The report will not include recommendations on policy issues such as funding, the creation of new programs or initiatives, or government organization.

BOX 1.2
Core and Emerging Areas Considered in This Report

Geodesy and geophysics

• Geodesy—the science of mathematically determining the size, shape, and orientation of the Earth, and the nature of its gravity field in four dimensions. It includes the development of highly precise positioning techniques, which enable monitoring of dynamic Earth phenomena such as ground subsidence and sea-level change. Related terms include surveying and navigation.

• Geophysics—the physics of the Earth and its environment in space, including the study of geodesy, geomagnetism and paleomagnetism, seismology, hydrology, space physics and aeronomy, tectonophysics, and atmospheric science.

Photogrammetry—the art, science, and technology of extracting reliable and accurate information about objects, phenomena, and environments from the processing of acquired imagery and other sensed data, both passively and actively, within a wide range of the electromagnetic energy spectrum.

Remote sensing—the science of measuring some property of an object or phenomenon by a sensor that is not in physical contact with the object or phenomenon under study.

Cartographic science—the discipline dealing with the conception, production, dissemination, and study of maps as both tangible and digital objects, and with their use and analysis.

Geographic Information Systems and geospatial analysis

• Geographic Information System—any system that captures, stores, analyzes, manages, and visualizes data that are linked to location.

• Geospatial analysis—the process of applying analytical techniques to geographically referenced data sets to extract or generate new geographical information or insight.

Geospatial Intelligence (GEOINT) fusion—the aggregation, integration, and conflation of geospatial data across time and space with the goal of removing the effects of data measurement systems and facilitating spatial analysis and synthesis across information sources.

Crowdsourcing—a process in which individuals gather and analyze information and complete tasks over the Internet, often using mobile devices such as cellular phones. Individuals with these devices form interactive, scalable sensor networks that enable professionals and the public to gather, analyze, share, and visualize local knowledge and observations and to collaborate on the design, assessment, and testing of devices and results. Related terms include volunteered geographic information, community remote sensing, and collective intelligence.

Human geography—the science of understanding, representing, and forecasting activities of individuals, groups, organizations, and the social networks to which they belong within a geotemporal context. It includes the creation of operational technologies based on societal, cultural, religious, tribal, historical, and linguistic knowledge; local economy and infrastructure; and knowledge about evolving threats within that geotemporal window. Related terms include cultural geography, spatial cultural intelligence, geo-enabled network analysis, and human terrain.

Visual analytics—the science of analytic reasoning, facilitated by interactive visual interfaces. The techniques are used to synthesize information and derive insight from massive, dynamic, ambiguous, and often conflicting data. Related terms include scientific visualization, information visualization, geovisualization, and visual reasoning.

Forecasting—an operational research technique used to anticipate outcomes, trends, or expected future behavior of a system using statistics and modeling. It is used as a basis for planning and decision making and is stated in less certain terms than a prediction. Related terms include prediction and anticipatory intelligence.

---

would continue at the current pace but would focus on the core and emerging areas. To estimate how many experts would likely be available in the future, the committee extrapolated the trend in the number of degrees conferred over the past 10 years to 2030.

The last two tasks address mechanisms for building knowledge and skills in the geospatial disciplines now and over the next 20 years. For Task 3, the committee described current government agency, university, professional society, and private company programs that

---

BOX 1.3
NGA Information Available for This Study

As an intelligence agency, little information on NGA's current activities, future plans, or the workforce needed to carry them out is publicly available. At the request of the committee, NGA provided the most essential information needed to carry out this study, including the following:

• NGA occupation descriptions (including education, knowledge, and skill requirements) for current scientist and analyst positions.
• The total number of scientists and analysts currently working in each geospatial intelligence occupation and the number hired each year over the past few years.
• The ages and highest degrees held by the current scientist and analyst workforce.
• The courses offered at the NGA College.
• The universities where NGA recruits or sends employees for training.
• The occupations tracked by the Bureau of Labor Statistics that are most relevant to NGA.

These data were provided in 2011; trends may have shifted significantly since the data were collected.

NGA did not provide strategic information, such as NGA hiring priorities, problems finding skills or expertise, or the basis for the NGA College curriculum. When such information was needed to support the analysis, the report states the assumptions made by the committee so readers can follow the reasoning.

BOX 1.4
NGA Scientist and Analyst Occupations

Geospatial intelligence is produced by scientists (including mathematicians) and analysts. Scientists are experts in a particular discipline, and they define NGA's research strategy, oversee scientific activities, apply new technologies, and develop expertise and tradecraft for the agency. Analysts acquire, process, and analyze data from government and commercial sources; ensure the quality, accuracy, and currency of geospatial information; populate databases; and produce information products for military and intelligence applications. NGA distinguishes more than 30 types of geospatial intelligence analysts, based on scientific discipline (e.g., geodetic earth science, nautical cartography, political geography) or function (e.g., data analysis, development of analysis methods, cross-disciplinary issues). Some analysts address agency-wide issues, such as developing multisource strategies to address intelligence problems, discovering and evaluating new open-source data, and tasking data collection systems. Descriptions of current NGA science and analyst occupations are given in Appendix B.

---

offer education or training in the disciplines, methods, and/or technologies underlying geospatial intelligence. Few of these programs are targeted to NGA's needs. For Task 4, the committee identified a short list of actions, of varying scope, that NGA can take to help build a skilled geospatial intelligence workforce in the future.

OVERVIEW OF THE NATIONAL GEOSPATIAL-INTELLIGENCE AGENCY

History

Military intelligence has always required mapping, cartographic analysis, and the collection of geographic information (Sweeney, 1924). The United States has supported mapping and charting for military intelligence purposes since 1804, when the Army's Lewis

and Clark expedition began exploring the Louisiana Territory (Table 1.1). Mapping and charting efforts advanced significantly during World War I, in part because of the extensive use of aerial photography for battlefield intelligence (e.g., MacLeod, 1919; Collier, 1994). In the World War II era, technological improvements in aircraft and cameras greatly expanded military applications of aerial photography, and maps began to be combined with analyzed imagery (e.g., Monmonier, 1985). The development of high-altitude aircraft in the mid-1950s enabled detailed maps of military bases, shipyards, and other strategic targets to be made, revealing, for example, the presence of Soviet medium-range ballistic missiles in Cuba in 1962 (e.g., Richelson, 1999). The advent of satellites in the late 1950s provided the capability to photograph the Earth, measure its physical properties, and accurately determine positions of objects on the surface (Table 1.1).

In the decades following World War II, the collection and handling of intelligence information from photogrammetry, geodesy, mapping, and charting became increasingly automated (Clarke, 2009). With automation came an improved ability to integrate different types of information and to carry out new types of analyses useful to decision makers, including time-

TABLE 1.1 Milestones in the Application of Core Areas to National Defense and Intelligence

Year	Event
1804	Lewis and Clark began to map and gather intelligence and other information on territory from St. Louis, Missouri, to the Pacific Ocean
1813	The Topographical Engineers began conducting surveys to facilitate the safe movement of troops for the War of 1812
1835	The Navy began to produce nautical charts and, 3 years later, to make astronomical observations
1889	The Army began to collect and compile information on geography and foreign forces, and to communicate it to military attachés during the Spanish-American War
1911	First photoreconnaissance flight. Aerial photography became a major contributor to battlefield intelligence during World War I
1922	First modern bathymetric chart, made using sounding data collected from a Navy ship
1928	The Army Air Corps began producing aeronautical charts
1941	Second World War aviation enabled photogrammetry, photo interpretation, and geodesy to replace field surveys
1953	Navy aircraft began measuring magnetic variations around the Earth; project U.S. Magnet continued until 1994
1956	High-altitude U-2 aircraft began to carry out manned reconnaissance missions, becoming the primary source for intelligence gathering over the Soviet Union and other denied areas
1960	Successful return of imagery from Corona, the first photoreconnaissance satellite system in the world
1960	World Geodetic System (WGS 60) defined an Earth-centered orientation system and formed the basis of current global positioning systems
1966	Launch of the Geodetic Earth Orbiting Satellite, the first dedicated satellite for geodetic studies
1974	First electronic dissemination of near-real-time, near-original-quality overhead imagery to support rapid targeting and assessment of strategic threats
1978	Launch of the first four Global Positioning System satellites, which enabled accurate measurements of position, velocity, and time
1994	Presidential directive PDD-23 directed the National Imagery and Mapping Agency to acquire commercial satellite data
1995	Unmanned aerial vehicles began taking streaming video during reconnaissance flights
2000	The Shuttle Radar Topography Mission began to acquire elevation data over about 80 percent of the Earth's surface using interferometric synthetic aperture radar
2005	Surface warships began to navigate using digital nautical charts
2006	First automatic construction of the three-dimensional world from diverse sources of photographs and images

SOURCES: Day et al. (1998); Snavley et al. (2006); Clarke (2013a); NGA historical reference chronology, <https://www1.nga.mil/About/OurHistory/Pages/default.aspx>.

space analysis and the evaluation of natural phenomena and human activities at the Earth's surface. NGA's current model for producing geospatial intelligence is illustrated in Figure 1.1 and an example of an information product is shown in Figure 1.2.

Organization

Through most of the 20th century, responsibility for specific aspects of mapping, charting, aerial photography, and eventually satellite reconnaissance was distributed among multiple defense and intelligence agencies and departments. In 1996, mapping, imagery acquisition and analysis, and intelligence production were brought together from the Defense Mapping Agency, the Central Imagery Office, and other imagery and mapping departments in a single agency—the National Imagery and Mapping Agency (NIMA).[1] NIMA's primary focus was on acquiring and providing imagery and maps to intelligence agencies. Increasing demands for speed, accuracy, and synthesis of geospatial information—especially since the September 2001 terrorist attacks in the United States—led to the concept of geospatial intelligence or GEOINT, the use of imagery and geospatial data to describe and depict features and activities and their location on the Earth. In 2003, the agency's name was changed to the National Geospatial-Intelligence Agency to emphasize its mission of producing geospatial intelligence.

NGA is part of the Department of Defense, and it is one of 16 federal agencies responsible for national intelligence. Its emphasis is on military and intelligence

---

[1] See NGA historical reference chronology, <https://www1.nga.mil/About/OurHistory/Pages/default.aspx>.

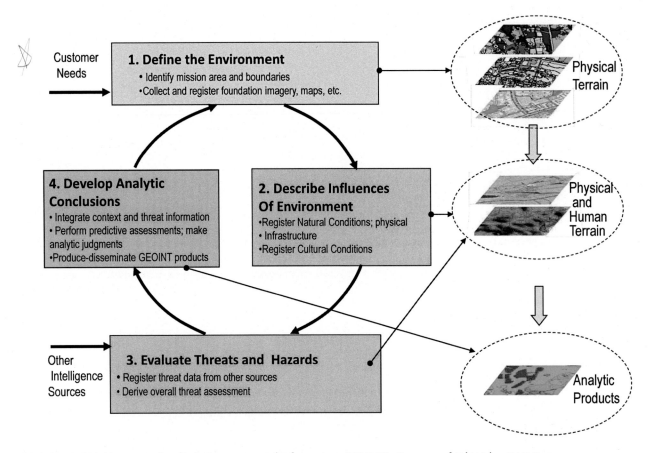

FIGURE 1.1 NGA's process for analyzing geospatial information. SOURCE: Courtesy of Ed Waltz, BAE Systems.

support in foreign countries, although humanitarian and disaster assistance, both at home and abroad, is a growing area of work for NGA. For example, NGA supported U.S. troops deployed to the Indian Ocean following the 2004 Sumatra earthquake and tsunami and provided imagery to U.S. and international relief organizations.[2] NGA also maintains the World Geodetic System, which is instrumental for both military and civil uses of the Global Positioning System.

NGA employs several thousand scientists and analysts, who acquire and analyze imagery and other geospatial information and deliver information products, services, and geospatial intelligence to policy makers, military decision makers, warfighters, and others. According to NGA, the largest fractions work on imagery analysis (about 40 percent), geospatial analysis (19 percent), and cartography (10 percent). Over the past few years, the agency has hired several hundred such

experts each year. A bachelor's degree or a combination of education and experience is preferred, although many NGA scientists and analysts have higher degrees. Additional training on sensors, geospatial analysis, and other subjects is provided by the National Geospatial-Intelligence College (hereafter referred to as the NGA College). NGA employees can also take classes at universities through the Vector Study Program.

ORGANIZATION OF THE REPORT

This report examines the supply of experts in 10 geospatial intelligence areas, gaps between the supply of experts and NGA's needs over the next 20 years, and ways to build necessary knowledge and skills. Chapter 2 characterizes the knowledge, skills, and academic programs in the five core areas that have historically underpinned geospatial intelligence, and Chapter 3 focuses on five emerging areas that could improve geospatial intelligence in the future. Chapter 4

---

[2] See NGA historical reference chronology, <https://www1.nga.mil/About/OurHistory/Pages/default.aspx>.

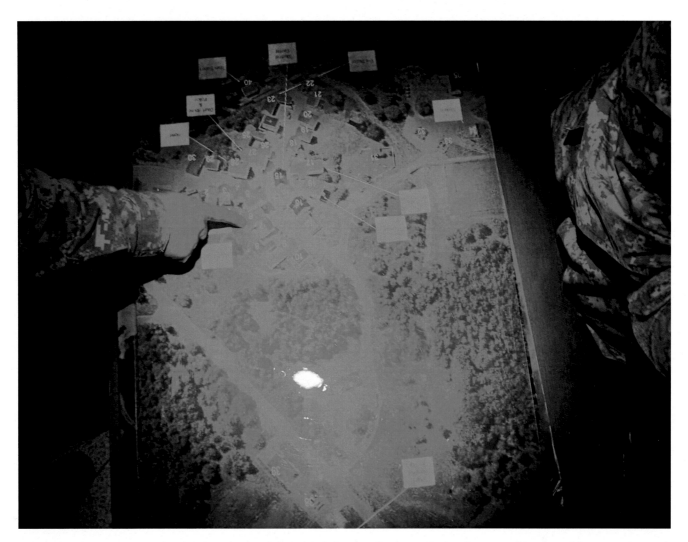

FIGURE 1.2 Army Research Laboratory's tactical digital hologram technology, which is being used by special forces in Iraq and Afghanistan. The unit has a three-dimensional holographic display that incorporates human intelligence, terrain, and imagery data. SOURCE: U.S. Army Research Laboratory.

assesses the current and future supply of geospatial intelligence expertise in these core and emerging areas, based on government statistics. Chapter 5 matches the supply of experts to NGA's needs, considering gaps in disciplinary knowledge and analytical skills, as well as where experts are recruited. Chapter 6 describes training programs in academia, government, industry, and professional societies that offer useful models for filling gaps in knowledge and skills. Potential mechanisms for building the supply of geospatial intelligence experts in the future are discussed in Chapter 7. Supporting material appears in the appendixes, including relevant university curricula and degree programs in the core and emerging areas (Appendix A), descriptions of scientist and analyst positions at NGA (Appendix B), and statistics on relevant degrees (Appendix C) and occupations (Appendix D). Biographical sketches of committee members are given in Appendix E, and a list of acronyms and abbreviations appears in Appendix F.

2

Core Areas of Geospatial Intelligence

Over the past several decades, the missions of agencies now represented in the National Geospatial-Intelligence Agency (NGA) have intersected with several academic fields, including geodesy, geophysics, cartographic science, geographic information science and spatial analysis, photogrammetry, and remote sensing. Advanced work in these fields depends on university research and curricula, the supply of graduate students, and technological advances. Agencies frequently sent employees to universities to gain specific expertise, for example to Ohio State University for geodesy (Cloud, 2000).

In recent years, many of these academic fields have become increasingly interdisciplinary and interrelated. For example, digital photogrammetry has so changed the field that its methods are barely distinguishable from remote sensing. Similarly, new labels such as geomatics have emerged, reflecting the overlap among surveying, photogrammetry, and geodesy. Few academic programs treat geographic information science, spatial analysis, and cartography as separate fields of study, but usually regard them as tracks or emphases within geography or another discipline. Professional organizations and academic journals reflect the interdisciplinary changes under way today. For example, mergers, name changes, and increasing overlap have characterized the professional organizations over the last decades (e.g., Ondrejka, 1997). This chapter examines how each of the core areas has evolved over time, the key concepts and methods that are currently taught, and the scope of existing education and professional preparation programs.

GEODESY AND GEOPHYSICS

Geodesy is the science of mathematically determining the size, shape, and orientation of the Earth and the nature of its gravity field in space over time. It includes the study of the Earth's motions in space, the establishment of spatial reference frames, the science and engineering of high-accuracy, high-precision positioning, and the monitoring of dynamic Earth phenomena, such as ground movements and changes in sea-level rise and ice sheets. Because much of contemporary geodesy makes use of satellite technology, such as the Global Positioning System (GPS), topics such as orbital mechanics and transatmospheric radio wave and light propagation also fall within its purview. Geophysics comprises a broad range of subdisciplines, including geodesy, geomagnetism and paleomagnetism, atmospheric science, hydrology, seismology, space physics and aeronomy, tectonophysics, and some ocean science. Given NGA's historical focus on geodesy, the following discussion concentrates on geodesy, touching on other subdisciplines of geophysics where appropriate.

Evolution

Geodesy is one of the oldest sciences whose study goes back to the ancient Greeks (e.g., Vaníček and Krakiwsky, 1986; Torge and Müller, 2012). The first attempt to accurately measure the Earth's size was made in the third century B.C. By measuring the lengths of shadows, Eratosthenes of Cyrene determined the Earth's circumference with an accuracy that would not be improved until the 17th century. The assumption that the Earth was a sphere was dispelled by Sir Isaac Newton. In the first edition of *Principia*, published in 1687, Newton postulated that the Earth was slightly ellipsoidal in shape, with the polar radius about 27 kilometers shorter than the equatorial radius. Refinements in field geodesy techniques slowly increased the accuracy of these estimates, but it was not until the dawn of the space age that knowledge of the Earth's size and shape improved significantly. Through the analysis of perturbations of satellite orbits, scientists first refined the ellipsoidal dimensions of the Earth and then discovered that the shape of the Earth, as represented by its gravity field, was much more complicated.

When geodesists talk about the shape of the Earth, what they actually mean is the shape of the equipotential surfaces of its gravity field. The equipotential surface that most closely approximates mean sea level is called the geoid. One of the major tasks of geodesy is to map the geoid as accurately as possible. An example of a highly accurate and precise geoid constructed using data from the Gravity field and steady-state Ocean Circulation Explorer (GOCE) satellite is shown in Figure 2.1 (Schiermeier, 2010; Floberghagen et al., 2011). Maps of the geoid provide information about the structure of the Earth's crust and upper mantle, plate tectonics, and sea-level change. The geoid is needed to accurately determine satellite orbits and the trajectories of ballistic missiles. It also finds everyday use as the surface from which orthometric heights, the heights usually found on topographic maps, are measured. Improved knowledge of the gravity field can also be combined with GPS and/or inertial navigation sensors to produce a more accurate navigation system than can be provided by GPS alone.

NGA's ongoing needs for geodesy stem primarily from work carried out by the former Defense Mapping Agency and include accurately and precisely determining the geoid, establishing accurate and precise coordinate systems (datums) and positions within them (e.g., World Geodetic System 1984; Merrigan et al., 2002), and relating different internationally used datums. In particular, NGA is responsible for supporting Department of Defense navigation systems, maintaining GPS fixed-site operations, and generating and distributing GPS precise ephemerides (Wiley et al., 2006).

Advances in geodesy are driven largely by continuing improvements to and expansion of space geodetic systems. New generations of GPS satellites are being deployed by the United States and several countries are developing global navigation satellite systems (GNSS), including the European Galileo, Chinese Compass, and Russian GLONASS systems. The use of GPS has become ubiquitous, with myriad civil and military applications. Improvements on the horizon include the development of less expensive and more accurate gravity gradiometry for determining the fine structure of the local gravity field and more accurate atomic clocks for measuring gravity and determining heights in the field.[1]

An important advance in geophysics that is relevant to NGA is the improvement in describing the Earth's ever-changing magnetic field. The National Geophysical Data Center's NGDC-720 model—compiled from satellite, ocean, aerial, and ground magnetic surveys—provides information on the field generated by magnetized rocks in the crust and upper mantle (Figure 2.2; Maus, 2010). This model is the first step toward producing a geomagnetic field model that would be useful for navigation.

Knowledge and Skills

Graduate study in geodesy encompasses the theory and modern practice of geodesy. Topics include the use of mathematical tools such as least-squares adjustment, Kalman filtering, and spectral analysis; the principles of gravity field theory and orbital mechanics; the propagation of electromagnetic waves; and the theory and operation of observing instruments such as GNSS receivers and inertial navigation systems. Modeling of observations to extract quantities of interest is a key technique learned by students. While course-only master's degrees are available at some universities, most graduate degrees in geodesy require completion of a research project, some of which involve substantial amounts of computer programming. Graduates may carry out or manage research, and traditionally have a master's or doctorate degree from a university specializing in geodesy and an undergraduate degree in a related field such as survey science, civil engineering, survey-

---

[1] Presentation by D. Smith, NOAA, to the NRC Workshop on New Research Directions for NGA, Washington, D.C., May 17-19, 2010.

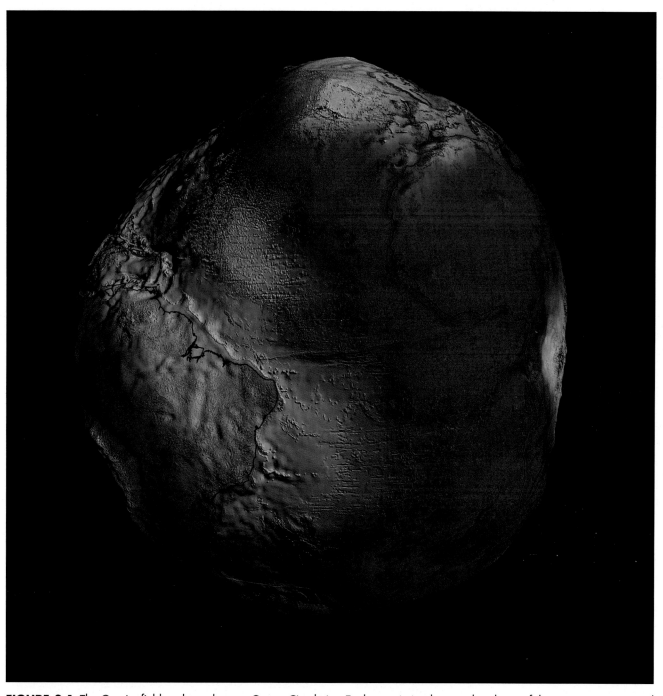

FIGURE 2.1 The Gravity field and steady-state Ocean Circulation Explorer mission has produced one of the most accurate geoid models to date. The deviations in height (−100 m to +100 m) from an ellipsoid are exaggerated 10,000 times in the image. The blue colors represent low values and the reds/yellows represent high values. SOURCE: ESA/HPF/DLR.

ing engineering, physics, astronomy, mathematics, or computer science.

The knowledge taught at the undergraduate level is similar in breadth, but less in depth than that taught at the graduate level. Courses include specialized mathematics such as adjustment calculus (least-squares analysis), geodetic coordinate systems and datums, the elements of the Earth's gravity field, and the basics of geodetic positioning techniques such as high-precision GPS surveying. Students should be well versed in the mathematical and physical principles underlying geodesy so that during their careers they can readily

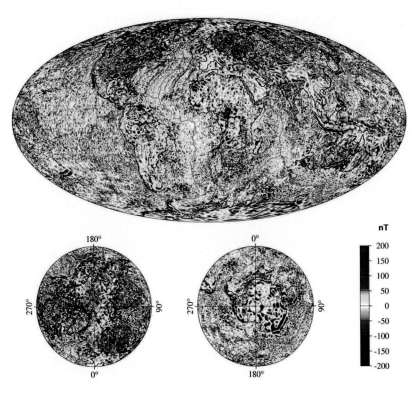

FIGURE 2.2 The downward-direction component of the crustal magnetic field, in nanoteslas, from the NGDC-720 model. The figure shows the magnetic potential, represented by spherical harmonic degree 16 to 720, which corresponds to the waveband of 2500 km to 56 km. SOURCE: National Geophysical Data Center.

adapt to advances in the field. Graduates with an undergraduate degree with geodesy as a major component commonly work as geodetic or surveying engineers, who design and supervise data collection activities, carry out routine analyses, and solve small problems of a theoretical nature.

A bachelor's degree in geophysics combines studies in geology and physics with mathematical training. Graduates commonly work as exploration geophysicists who prospect for oil, gas, or minerals; or as environmental geophysicists who assess soil and rock properties for various applications. A graduate degree in geophysics, preferably a doctorate, is required for research. Graduate-level knowledge and skills acquired in geophysics programs mirrors that in geodesy programs, with some overlap in subject areas. Additional topics of study include seismology and the structure and evolution of the Earth, including plate tectonics, the theory and measurement of the Earth's magnetic field, and space physics, including the nature of the ionosphere and magnetosphere and the phenomena of space weather and its impact on modern technological systems.

Education and Professional Preparation Programs

At the undergraduate level, geodesy is primarily taught in geomatics programs (Box 2.1), typically in a geomatics or surveying engineering department or as an option in a civil engineering department, and sometimes in other departments (e.g., earth science, aerospace engineering, forestry). A few geography programs teach geomatics, but there is typically little geodesy content.

Only a handful of undergraduate geomatics programs (e.g., University of Florida, Texas A&M University, Corpus Christi) currently exist in the United States. More existed in the past[2] but were terminated because of reduced demand or a change in institutional priorities. In some cases, the associated graduate program survived. At the graduate level, geodesy is taught in geomatics, geophysics, earth science, planetary

---

[2] In the late 1970s, 13 schools in the United States offered 4-year bachelor's programs in surveying or geodetic science, 8 offered a master's degree in surveying, and 6 offered a Ph.D. in surveying and/or geodesy (NRC, 1978).

BOX 2.1
Geomatics

Geodesy provides the scientific underpinning for geomatics, a relatively new term used to describe the science, engineering, and art involved in collecting and managing geographically referenced information. A number of government agencies, private companies, and academic institutions have embraced this term as a replacement for "surveying and mapping," which no longer adequately describes the full spectrum of position-related tasks carried out by professionals in the field. Geomatics covers activities ranging from the acquisition and analysis of site-specific spatial data in engineering and development surveys to cadastral and hydrographic surveying to the application of GIS and remote sensing technologies in environmental and land use management.

science, or engineering (primarily instrumentation-related) departments. Again, only a few such degree programs (e.g., Massachusetts Institute of Technology, Ohio State University) currently exist in the United States. Notable examples of U.S. universities currently offering an undergraduate degree in geomatics or a graduate degree in geodesy are listed in Table A.1 in Appendix A.

Some 2-year colleges and associate degree programs in universities offer programs in surveying or geomatics technology, which provide basic instruction in the principles of geodesy, including coordinate systems and the use of GPS. There are many such colleges across the United States, whose primary purpose is to produce surveying and mapping technicians. Examples include the Geomatics Technology Program at Greenville Technical College (South Carolina) and the Engineering Technology Program at Alfred State College (New York).

Course-only master's degrees offered by some of the institutions mentioned in Appendix A allow entry into some geodesy-related jobs. Some professional-level education in geodesy is also available through continuing education programs and short courses offered by diverse organizations, such as the National Geodetic Survey, NavtechGPS, the Institute of Navigation, Pennsylvania State University, and the Michigan Technical University.

Undergraduate degrees or specialization in geophysics are available at a number of universities in departments of physics, earth and planetary sciences, and geology and geophysics (e.g., Stanford University, Harvard University; see Table A.1 in Appendix A). Many universities also offer master's and doctorate degree programs in geophysics, including the California Institute of Technology and the Massachusetts Institute of Technology.

PHOTOGRAMMETRY

The term photogrammetry is derived from three Greek words: photos or light; gramma, meaning something drawn or written; and metron or to measure. Together the words mean to measure graphically by means of light. Photogrammetry is concerned with observing and measuring physical objects and phenomena from a medium such as film (Mikhail et al., 2001). Whereas photographs were the primary medium used in the early decades of the discipline, many more sensing systems are now available, including radar, sonar, and lidar, which operate in different parts of the electromagnetic radiation spectrum than the visual band (Kraus, 2004). Moreover, while most early activities involved photography from manned aircraft, platforms have since expanded to unmanned vehicles, satellites, and handheld and industrial sensors. Construction of a mathematical model describing the relationship between the image and the object or environment sensed, called the sensor model, is fundamental to all activities of photogrammetry (McGlone et al., 2004). Given these changes in the field, photogrammetry is now defined as the art, science, and technology of extracting reliable and accurate information about objects, phenomena, and environments from acquired imagery and other sensed data, both passively and actively, within a wide range of the electromagnetic energy spectrum. Although its emphasis is on metric rather than thematic content, imagery interpretation, identification of targets, and image manipulation and analysis are required to support most photogrammetric operations.

In photogrammetry, the Earth's terrain is imaged using overlapping images (photographs) taken from aircraft or hand-held cameras, linear scans of an area from a satellite (Figure 2.3), or data from active sensors, such as radar, sonar, and laser scanners. A single image, which is a two-dimensional recording of the three-dimensional (3D) world, is not sufficient to determine all three ground coordinates of any target point. Unless one of the three coordinates is known, such as the elevation from a digital elevation model, two or more images are required to accurately recover all three dimensions (Figure 2.4). Imagery, sensor and platform parameters, and metadata such as that from GPS and INS (inertial navigation system) are used in the photogrammetric exploitation.

Most photogrammetric activities deal with cameras and sensors that are carefully built and calibrated to allow direct micrometer-level measurements. However, an important branch of photogrammetry deals with less sophisticated instruments, such as those found on mobile phones, which require careful modeling and often self-calibration. This branch is gaining importance as the availability of imagery from nonmetric cameras grows.

Many digital photogrammetric workstations enable the overlap area of two images to be viewed stereoscopically. Automated algorithms are commonly used to extract 3D features with high accuracy. Frequently, however, human judgment is required to edit, or sometimes to override, the results from such algorithms.

Evolution

Photogrammetry began as a branch of surveying and was used for constructing topographic maps and for military mapping. It is still sometimes taught in surveying departments. Technological advances in surveying, the growth of photogrammetry, and the inclusion of related fields, such as geodesy, remote sensing (Box 2.2), cartography, and GIS, made the title "surveying" or "surveying engineering" inadequate for a department. The name geomatics or geomatics engineering was introduced to better capture this range of activities (see Box 2.1). At present, photogrammetry is taught in geomatics departments, as well as in other departments, such as geography and forestry.

Photogrammetry has gone through three stages of development: analog, analytical, and digital (Blachut and Burkhardt, 1989). Analog instruments were built to optomechanically simulate the geometry of passive imaging and to allow the extraction, mostly graphically, of information in the form of maps and other media. As computers became available, mathematical models of sensing were developed and algorithms were imple-

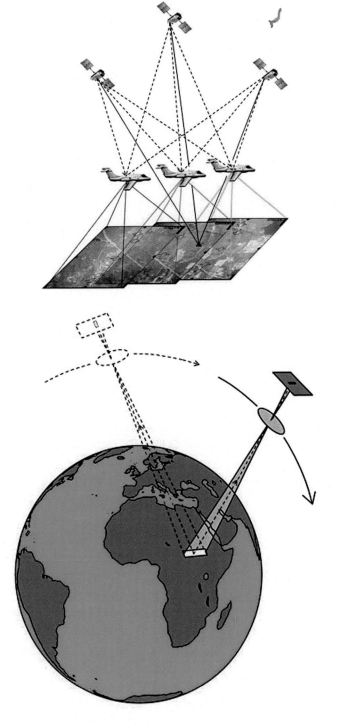

FIGURE 2.3 Accurate photogrammetric reconstruction of the imaged terrain requires overlapping images and metadata. (*Top*) Overlapping frame images produced by an aircraft; metadata include aircraft location determined from a constellation of GPS satellites, orientation determined from an inertial navigation system, and/or GPS-determined ground control (red triangle). (*Bottom*) Overlapping images produced by linear array scans from a satellite.

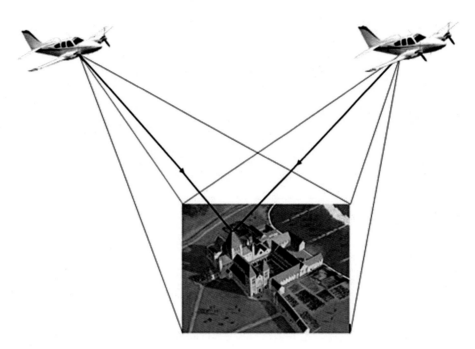

FIGURE 2.4 Recovery of three-dimensional target points requires at least two overlapping images, which is the basis for accurate stereo photogrammetry.

BOX 2.2
Photogrammetry and Remote Sensing

Both photogrammetry and remote sensing originated in aerial photography. Before it was called remote sensing, this field focused on identifying what is recorded in a photograph. By contrast, photogrammetry was concerned with where the recorded objects are in geographic space. Therefore, photogrammetry required more information about the photography, such as the camera characteristics (e.g., focal length, lens distortion) and aircraft trajectory (e.g., altitude, camera attitude). Airphoto interpretation requires less precise knowledge of the geometry of the photographs; it may suffice to know the approximate scale.

The term remote sensing was introduced with the advent of systems that sense in several regions of the electromagnetic spectrum. For decades, remote sensing involved many of the same activities as photogrammetry at a coarser resolution, but contemporary remote sensing can image at resolutions equivalent to those used in photogrammetry. What used to be almost entirely done by a human—the interpretation of photographs—has now been replaced by sophisticated algorithms based on mathematical pattern recognition and machine learning. Nevertheless, the fundamental tasks of the disciplines remain essentially the same. In photogrammetry, one deals with the rigorous mathematical modeling of the relationship between the sensed object and its representation by the sensor. Through such models, various types of information can be extracted from the imagery, such as precise positions, relative locations, dimensions, sizes, shapes, and all types of features. High accuracy is critical. For example, accurate modeling is used in multiband registration of multispectral imagery. In remote sensing, the goal is usually to transform an image so that it is suitable for mapping some property of the Earth surface synoptically, such as soil moisture or land cover.

---

mented primarily in batch mode. The transition from analog to analytical was epitomized by the introduction of the analytical plotter in 1961, which incorporated a dedicated computer. The development of the digital photogrammetric workstation ushered in the stage of digital photogrammetry.

Advances in optics, electronics, imaging, video, and computers during the past three decades have led to significant changes in photogrammetry. Film is being replaced by digital imagery, including imagery from active sensors, such as radar and, more recently, lidar.[3] The operational environment and the variety of activities and products have also changed dramatically. The range of products has broadened beyond image products (e.g., single, rectified, and orthorectified images; mosaics; radar products) to point and line products (e.g., targets, digital surface models, digital elevation models, point clouds, vectors) to relative information products (e.g., lengths, differences, areas, surfaces, volumes) to textured 3D models. Photogrammetry products now provide the base information for many geographic information systems (GIS). Finally, many processes are being automated, allowing near-real-time applications. The next phase may well be called on-demand photogrammetry, with many activities based online. It is likely that processing will be pushed upstream toward the acquisition platform, making it possible to obtain information products, rather than data, from an airborne or satellite sensor. Direct 3D imaging may be imminent. Photogrammetry will likely continue to play a significant role in ascertaining precision and accuracy of geospatial information, and to contribute to the complex problem of fusing imagery with other data.

Knowledge and Skills

Photogrammetry classes are taught in undergraduate programs in surveying, surveying engineering, geomatics, or geomatics engineering, but none of these programs in the United States offer a bachelor's degree in photogrammetry. The graduates of such programs may be employed in mapping firms, particularly if they took an extra elective course in photogrammetry. They would know how aerial photography and other imagery is acquired and how to use it in stereoscopic processing systems to extract various types of mapping information. It is likely that they would receive significant on-the-job training by seniors in their firm.

The individuals who obtain a master's degree in photogrammetry gain much more knowledge based on a strong mathematical foundation. Such photogrammetrists or photogrammetric engineers design algorithms to exploit various types of imagery.

They understand the different platforms and have a command of the techniques of least-squares adjustment and estimation from redundant measurements. Photogrammetric scientists usually have a doctorate and are capable of supervising or carrying out research and modeling the various complex imaging systems. They conceive of novel approaches and ways to deal with technological advances, whether in new sensors, new modes of image acquisition from orbital platforms or aircraft, or in the integration and fusion of information from multiple sources.

Education and Professional Preparation Programs

Education programs in photogrammetry (e.g., Ohio State University, Cornell, Purdue University) flourished in the early and mid-1960s. At the time, photogrammetry was being used extensively by the Defense Mapping Agency, the U.S. Geological Survey, the U.S. Coast and Geodetic Survey, the military services, and the intelligence community. Demand for training was high, and these organizations sent significant numbers of employees to universities under programs such as the Long Term Full Time Training (LTFTT) program. By the late 1980s and early 1990s, more than 25 photogrammetry programs were offering both master's and doctorate degrees in the field. At the undergraduate level, photogrammetry was introduced as a small part of undergraduate courses in surveying and mapping. In the 1980s and 1990s, several institutions (e.g., Ferris State, California State University, Fresno) offered lower-level photogrammetry courses as part of their undergraduate bachelor's programs in forestry, geography, civil engineering, construction engineering, surveying engineering, and, most recently, geomatics. About that time, the Defense Mapping Agency embarked on a modernization program (MARK 85 and MARK 90) to convert to digital imagery and move toward automation. The agency's focus on professional development shifted from learning fundamental principles to mastering skills to run software for photogrammetry applications. By the mid-1990s, the number of students taking classes through the LTFTT program and its successor Vector Study Program began to decrease significantly, and the decline in enrollment reduced support for educational programs offering a substantial emphasis in photogrammetry.

---

[3] Although terms such as radargrammetry and lidargrammetry are sometimes used to emphasize the type of sensor data being analyzed, the fundamentals of photogrammetry apply to all types of sensor data.

At present, only a handful of programs in photogrammetry exist in the United States (see Table A.2 in Appendix A). A few, such as those at Ohio State University and Purdue University, are top tier, yet are struggling to survive. Retiring faculty are not being replaced, and the number of faculty will soon decline below the critical mass needed to sustain these programs. Some 2-year technology programs, such as in surveying or construction technology, offer hands-on training using photogrammetric instruments to compile data. Most of these provide some photogrammetric skills but lack the rigorous mathematical basis of photogrammetry programs in 4-year colleges.

Outside of formal academic education, employers often provide in-house training, and some educational institutions and professional societies offer short courses ranging from a half day to a full week. The American Society for Photogrammetry and Remote Sensing regularly devotes a day or more to concurrent half-day or full-day short courses on specific topics in conjunction with its annual and semiannual meetings. Most of those who take these courses are employees seeking professional development.

REMOTE SENSING

Remote sensing is the science of measuring some property of an object or phenomenon by a sensor that is not in physical contact with the object or phenomenon under study (Colwell, 1983). Remote sensing requires a platform (e.g., aircraft, satellite), a sensor system (e.g., digital camera, multispectral scanner, radar), and the ability to interpret the data using analog and/or digital image processing.

Evolution

Remote sensing originated in aerial photography. The first aerial photograph was taken from a tethered balloon in 1858. The use of aerial photography during World War I and World War II helped drive the development of improved cameras, films, filtration, and visual image interpretation techniques. During the late 1940s, 1950s, and early 1960s, new active sensor systems (e.g., radar) and passive sensor systems (e.g., thermal infrared) were developed that recorded electromagnetic energy beyond the visible and near-infrared

part of the spectrum. Scientists at the Office of Naval Research coined the term remote sensing to more accurately encompass the nature of the sensors that recorded energy beyond the optical region (Jensen, 2007).

Digital image processing originated in early spy satellite programs, such as Corona and the Satellite and Missile Observation System, and was further developed after the National Aeronautics and Space Administration's (NASA's) 1972 launch of the Earth Resource Technology Satellite (later renamed Landsat) with its Multispectral Scanner System (Estes and Jensen, 1998). The first commercial satellite with pointable multispectral linear array sensor technology was launched by SPOT Image, Inc., in 1986. Subsequent satellites launched by NASA and the private sector have placed several sensor systems with high spatial resolution in orbit, including IKONOS-2 (1×1 m panchromatic and 4×4 m multispectral) in 1999, and satellites launched by GeoEye, Inc. and DigitalGlobe, Inc. (e.g., 51×51 cm panchromatic) from 2000 to 2010. Much of the imagery collected by these companies is used for national intelligence purposes in NGA programs such as ClearView and ExtendedView.

Modern remote sensing science focuses on the extraction of accurate information from remote sensor data. The remote sensing process used to extract information (Figure 2.5) generally involves (1) a clear statement of the problem and the information required, (2) collection of the in situ and remote sensing data to address the problem, (3) transformation of the remote sensing data into information using analog and digital image processing techniques, and (4) accuracy assessment and presentation of the remote sensing-derived information to make informed decisions (Jensen, 2005; Lillesand et al., 2008; Jensen and Jensen, 2012).

State-of-the-art remote sensing instruments include analog and digital frame cameras, multispectral and hyperspectral sensors based on scanning or linear/area arrays, thermal infrared detectors, active microwave radar (single frequency-single polarization, polarimetric, interferometric, and ground penetrating radar), passive microwave detectors, lidar, and sonar. Selected methods for collecting optical analog and digital aerial photography, multispectral imagery, hyperspectral imagery, and lidar data are shown in Figure 2.6. Lidar imagery is increasingly being used to produce digital surface models, which include veg-

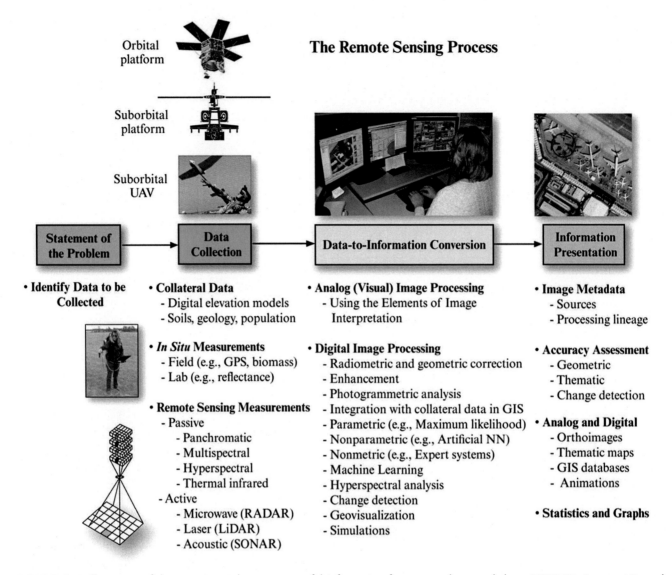

FIGURE 2.5 Illustration of the process used to extract useful information from remotely sensed data. SOURCE: Jensen, J.R. and R.R. Jensen, *Introductory Geographic Information Systems*, ©2013. Printed and electronically reproduced by permission of Pearson Education, Inc., Upper Saddle River, New Jersey.

etation structure and buildings information, and bare-earth digital terrain models (NRC, 2007; Renslow, 2012).

Airborne and satellite remote sensing systems can now function as part of a sensor web to monitor and explore environments (Delin and Jackson, 2001). Unlike sensor networks, which merely collect data, each sensor in a sensor web has its own microprocessor and can react and modify its behavior based on data collected by other sensors in the web (Delin, 2005). The individual sensors can be fixed or mobile and can be deployed in the air, in space, and/or on the ground. A few of the sensors can be configured to transmit infor-

mation beyond the local sensor web, which is useful for obtaining situational awareness (Delin and Small, 2009). Remote sensing systems are likely to find even greater application in the future when used in conjunction with other sensors in a sensor web environment.

Knowledge and Skills

Although curricula for educating remote sensing scientists and professionals have been developed,[4] they have not been widely adopted. Ideally, undergraduate

---

[4] See <http://rscc.umn.edu/>.

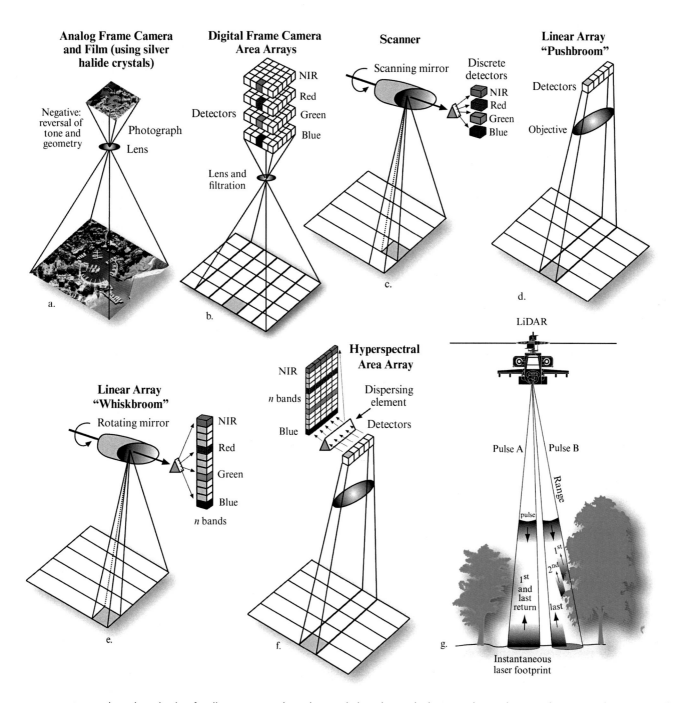

FIGURE 2.6 Selected methods of collecting optical analog and digital aerial photography, multispectral imagery, hyperspectral imagery, and lidar data. SOURCE: Jensen, J.R., *Remote Sensing of the Environment: An Earth Resource Perspective*, 2nd, ©2007. Printed and electronically reproduced by permission of Pearson Education, Inc., Upper Saddle River, New Jersey.

and graduate students specializing in remote sensing at universities are well versed in a discipline (e.g., forestry, civil engineering, geography, geology); understand how electromagnetic energy interacts with the atmosphere and various kinds of targets; are trained in statistics, mathematics, and programming; and know how to use a GIS (Foresman et al., 1997). Remote sensing scientists and professionals must be able to analyze digital remote sensor data using a diverse array of digital image processing techniques, such as radiometric and geometric preprocessing, enhancement (e.g., image fusion, filtering), classification (e.g., machine learning, object-oriented image segmentation, support vector machines), change detection and animation, and the integration of digital remote sensor data with other geospatial data (e.g., soil, elevation, slope) using a GIS (Jensen et al., 2009). Skills are also needed to interpret real-time video imagery collected from satellite, sub-orbital, and unmanned aerial vehicles.

Education and Professional Preparation Programs

There are no departments of remote sensing in U.S. universities (Mondello et al., 2006, 2008). Instead, a variety of departments offer degree tracks in remote sensing as part of a degree in other fields, including

- geography (all types of remote sensing),
- natural resources/environmental science (all types of remote sensing),
- engineering (sensor system design and all types of remote sensing),
- geomatics (all types of remote sensing),
- geology/geoscience (all types of remote sensing and ground penetrating radar),
- forestry (all types of remote sensing, but especially lidar),
- anthropology (especially the use of aerial photography and ground penetrating radar), and
- marine science (especially the use of aerial photography and sonar).

Few of these programs offer lidar courses; most lidar instruction takes place within other remote sensing courses.

Dozens of departments at 4-year universities offer degree tracks in remote sensing. A selected list of departments with a remote sensing-related concentration, track, or degree appears in Table A.3 in Appendix A. Geography programs offer more remote sensing courses and grant more degrees specializing in all types of remote sensing than any other discipline.

As far as can be determined, few remote sensing courses are offered at 2-year colleges, and no degrees are granted with a specialization in remote sensing. Remote sensing education is also available through workshops and webinars organized by professional societies and online instruction and degrees offered by universities.

CARTOGRAPHIC SCIENCE

Cartography focuses on the application of mathematical, statistical, and graphical techniques to the science of mapping. The discipline deals with theory and techniques for understanding the creation of maps and their use for positioning, navigation, and spatial reasoning. Components of the discipline include the principles of information design for spatial data, the impact of scale and resolution, and map projections (Slocum et al., 2009). Themes often analyzed include evaluation of design parameters—especially those involved with symbol appearance, hierarchy, and placement—and assessment of visual effectiveness. Other topics emphasized include transformations and algorithms, data precision, and data quality and uncertainty. Cartography also focuses on automation in the production, interpretation, and analysis of map displays in paper, digital, mobile device, and online form.

Among the key tasks that fall within cartography at NGA are maintaining geographic names data, producing standard map coverage for areas outside the United States, and nautical and aeronautical charting (e.g., Figure 2.7). The operational demands of the armed services for digital versions of standard maps and charts have expanded with the increased availability of automated navigation systems.

Evolution

The roots of cartography are positioned in geodesy and surveying, in exploration for minerals and natural resources, in maritime trade, and in sketching and lithographic renderings of landscapes by geologists and geographers. The formal discipline of cartography dates back to the late 1700s, when William Playfair began mapping thematic information on demographic, health, and socioeconomic characteristics. Military and strategic applications, particularly navigation and ballistics, have driven many of the major advances in cartography. Improvements in printing, flight, plastics, and electronics supported cartographic production, distribution, preservation, spatial registration, and automation.

The end of World War II created a surplus of trained geographers who moved from military intelligence to academic positions. During the 1970s and

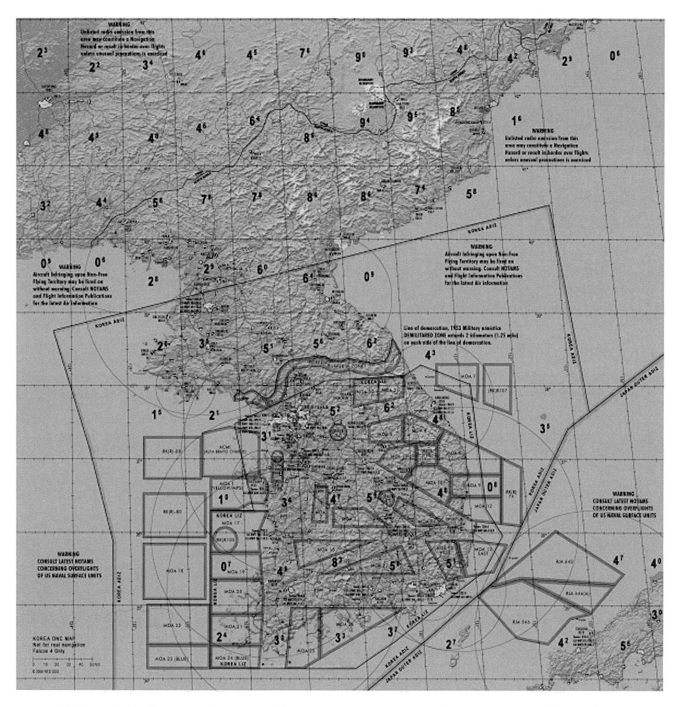

FIGURE 2.7 NGA digital operational navigational chart covering the Korean peninsula at 1:1M scale, displayed in the Falconview software. SOURCE: Clarke (2013b).

1980s, graduate programs specializing in cartography began to emerge at about a dozen universities. Beginning in the early 1980s, GIS began to flourish, largely due to the decision to automate the U.S. Decennial Census and map production at the U.S. Geological Survey (McMaster and McMaster, 2002). Demands for personnel trained in processing spatial information increased. In response, the emphasis of university curricula shifted from cartography to geographic information science (Box 2.3).

Cartographic skills in information design, data modeling, map projections, coordinate systems, and

---

BOX 2.3
Geographic Information Science

Geographic information science is a term coined in a seminal article by Michael F. Goodchild (1992) to encompass the scientific questions that arise from geographic information, including both research about GIS that would lead eventually to improvements in the technology and research with GIS that would exploit the technology in the advancement of science (Goodchild, 2006). As such, geographic information science includes aspects of cartography, computer science, spatial statistics, cognitive science, and other fields that pertain to the analysis of spatial information, as well as societal and ethical questions raised by the use of GIS (e.g., issues of privacy).

---

statistical analysis for mapping remain an important foundation for many tasks in geospatial intelligence. For example, an ability to create and interpret interactive and real-time graphical displays of geographic spaces (e.g., streaming video footage of enemy terrain) or of statistical information spaces (e.g., statistical clusters of demographic, economic, political, and religious characteristics) could help identify latent or developing terrorist cells. Skills required for nautical charting include a working knowledge of calculus, solid programming skills, and expertise in converting among international geodetic datums and spheroids. A nautical charting specialist must also be able to compile information from various sources and establish a statistical confidence interval for each information source and to quantify data reliability.

An emerging area of cartography, which addresses the design and analysis of statistical information displays, has been called geovisualization (Dykes et al., 2005) or geographic statistical visualization (Wang et al., 2002). Whereas scientific visualization is focused on realistic renderings of surfaces, solids, and landscapes using computer graphics (McCormick et al., 1987; Card et al., 1999), geovisualization emphasizes information design that links geographic and statistical patterns (e.g., Figure 2.8). The primary purpose of geovisualization is to illustrate spatial information in ways that enable understanding for decision making and knowledge construction (MacEachren et al., 2004). Its practical applications include urban and strategic

planning, resource exploration in hostile or inaccessible environments, modeling complex environmental scenarios, and tracking the spread of disease. A superset of this area, called visual analytics, is described in Chapter 3.

The transition from traditional cartography to geographic information science in universities has changed the mix of knowledge and skills being taught. Basic cartographic skills remain a prerequisite to geographic information science training, which requires understanding of projections, scale, and resolution. Virtually all GIS textbooks include basic information on cartographic scale, map projections, coordinate systems, and the size and shape of the Earth. Knowledge about the principles of graphic display has been deemphasized in most curricula, even though map displays in GIS environments are often created by analysts and are subject to misinterpretation. The traditional cartographic training in map production has been replaced by training in cartography, in detection and identification of spatial relationships, in spatial data modeling, and in the application of mapping to spatial pattern analysis. Many curricula have also incorporated coursework to train students in the use of GIS. In the past decade, most curricula have introduced coursework in software programming, database management, and web-based mapping and data delivery.

The minimum cartographic skills needed for professional cartographers include a demonstrated ability to work with basic descriptive and inferential statistics; an ability to program in C++, Java, or a scripting language such as Python; understanding of the principles of information design (Bertin, 1967); and a working knowledge of current online and archived data sources and software for their display. Professional cartographers are capable of handling large data sets, of undertaking basic and advanced statistical analysis (difference of means, correlation, regression, interpolation) in a commercial software environment, of interpreting spatial patterns in data, and of representing these patterns effectively on charts and map displays.

Cartographic skills used in the subdiscipline of geovisualization include map animation, geographic data exploration, interactive mapping, uncertainty visualization, mapping virtual environments, and collaborative geovisualization (Slocum et al., 2009).

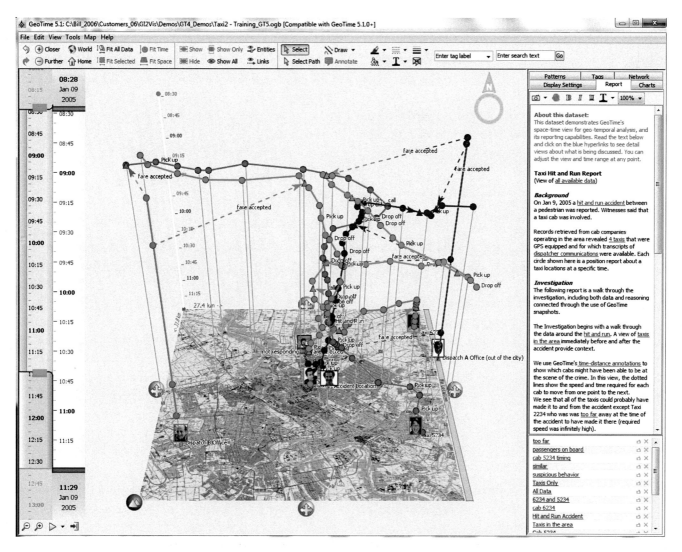

FIGURE 2.8 Example of a geovisualization technique that allows the display of events unfolding over time (vertical axis) and space (map). SOURCE: GeoTime is a registered trademark of Oculus Info Inc. Image used by permission of Oculus Info Inc.

Education and Professional Preparation Programs

A few dozen academic geography departments in the United States offer a degree track or concentration or a certificate with cartography or mapping in the title of the degree or certificate (see examples in Table A.4 in Appendix A). Students enrolled in these degree tracks or certificate programs are commonly required to take two or more courses related to cartography and mapping, as well as a course in basic statistics. At present, the most diverse undergraduate curriculum in cartography is offered by the University of Wisconsin. Strong graduate programs in cartography are harder to identify since so many graduate curricula have been folded into geographic information science work.

There are four major career paths in cartography: (1) information design, which focuses on design and graphic representation for topographic, reference (atlas), or thematic mapping; (2) GIS analysis (see below); (3) visual analytics (see Chapter 3); or (4) production cartography, which focuses on printing and reproduction. As the demand for production cartographers declines, fewer programs offer a primary or even a secondary focus on printing and reproduction. The demand for web, mobile, and online map production continues to grow, however. It is possible to take web or mobile coursework at some U.S. colleges and universities, but presently there are no certificate or degree programs in these topics. There is a demand for professionally trained cartographic designers to produce atlas and topographic map designs, and undergraduate training in this area can be found at several universities, such as Oregon State University, Pennsylvania State University, and Salem State University.

A number of 2-year colleges offer coursework in cartography and geographic information science. The shorter time required to complete a degree coupled with smaller class sizes (relative to larger universities) provides an environment conducive to hands-on training, which is essential preparation for good cartographic practice. Laboratory assignments, courses including practical work, and semester projects which are offered in 2-year colleges may not be offered until junior or senior year at universities, simply due to the size of the student population. The disadvantage of the 2-year college curricula, however, is that less attention is paid to computational and statistical skills, mostly due to the shortened time span.

Many universities offer professional preparation in geographic information science, and, in the best programs, cartography courses are a prerequisite to GIS courses. Most professional preparation in cartography that is relevant to geospatial intelligence focuses on GIS analysis or visual analytics. GIS analysts with cartographic training have a better understanding of projections and scale dependence. Important spatial patterns may be evident only in data within specific scale ranges, and cartographers are trained to be sensitive to relationships between spatial process and spatial or temporal resolution. Visual analytics experts with cartographic training bring an understanding of spatial relationships (also known as spatial thinking or reasoning; see NRC, 2006), which is endemic to geographic training. Career preparation in cartography also includes training in basic statistics, which is necessary for exploring and interpreting spatial patterns. Geovisualization shows great promise for integrating geographic, cognitive, and statistical skill sets for creation, analysis, and interpretation of geographical and statistical information displays, all of which are valuable for military intelligence.

GEOGRAPHIC INFORMATION SYSTEMS AND GEOSPATIAL ANALYSIS

Geographic information systems are computer-based systems that deal with the capture, storage, representation, visualization, and analysis of information that pertains to a particular location on the Earth's surface. Geospatial analysis emphasizes the extraction of information, insight, and knowledge from the GIS through the application of a wide range of analytical techniques, including visualization, data exploration, statistical and econometric modeling, process modeling, and optimization (e.g., Figure 2.9).

Evolution

GIS evolved to a reasonably well-defined discipline from a variety or origins, including cartography, land management, computer science, urban planning, and landscape architecture. Geospatial analysis has its roots in analytical cartography, the quantitative approach to geography pioneered at the University of Washington, and the development of quantitative spatial methods in regional science and operations research dating back to the early 1960s. Its early scope is represented by the classic book *Spatial Analysis: A Reader in Statistical Geography* (Berry and Marble, 1968). While often identified with spatial statistics, geospatial analysis encompasses a range of techniques from visualization to optimization. The need to develop analytical techniques to accompany the technology of GIS was raised by a number of scholars in the late 1980s and early 1990s (e.g., Goodchild, 1987; Anselin and Getis, 1992; Goodchild et al., 1992). Compilations of early progress in GIS and geospatial analysis appear in Fotheringham and Rogerson (1994) and Fischer and Getis (1997), and comprehensive overviews of the state of the art are provided in Fotheringham and Rogerson (2009), Fischer and Getis (2010), and de Smith et al. (2010).

Both GIS and geospatial analysis are changing rapidly as a result of the creation of Google Earth and similar services, the ready availability of technology to support location-based services and analysis, and the use of the Internet as cyberinfrastructure. These technological changes challenge the traditional model of an industry dominated by the products of a small number of vendors. Increasingly, GIS is offered as a web service and credible open-source competitors to the commercial platforms are appearing, supported by open standards developed by organizations such as the Open Geospatial Consortium. This development has significantly democratized access to geographic information,

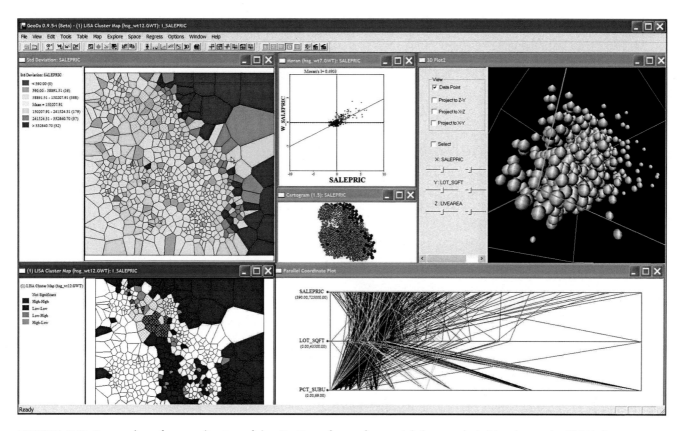

FIGURE 2.9 Screen shot of an application of the GeoDa software for spatial data analysis (Anselin et al., 2006) illustrating an exploration of spatial patterns of house prices in Seattle, Washington. The different graphs and maps are dynamically linked in the sense that selected observations (highlighted in yellow) are simultaneously selected in all windows. SOURCE: Anselin (2005).

which relies increasingly on a web browser to query, analyze, and visualize spatial data. Crowdsourcing is becoming more important, changing the role of traditional data providers, and the notion of cyberGIS (extensions of cyberinfrastructure frameworks that account for the special characteristics of geospatial data and geospatial analytical operations, e.g., Wang, 2010) is around the corner. Research in geospatial analysis is embracing the study of space-time dynamics associated with both human and physical phenomena, increasingly supported by massive quantities of data. This new direction requires new conceptual frameworks, methods, and computational techniques and is driving a rapidly evolving state of the art.

Knowledge and Skills

GIS and geospatial analysis are taught in undergraduate and graduate curricula in a wide range of university programs, such as geography, urban planning, landscape architecture, ecology, anthropology, and civil engineering. The core curriculum for GIS education is laid out in the "Body of Knowledge" (DiBiase et al., 2006), which is used by many higher education institutions to help structure GIS offerings.[5] The core curriculum outlines a range of necessary knowledge and skills, including a solid foundation in cartography, information systems, computer science, geocomputation, statistics, and operations research. Most university programs include coursework in a subset of these skills, but few deliver the full range of skills.

Education and Professional Preparation Programs

GIS educational programs and their degree of technical sophistication vary widely and range from community college training to undergraduate and

_____

[5] Community input is currently being gathered for the second edition of the *Body of Knowledge*.

graduate certificates to master's and professional master's programs. There are some 189 GIS degree programs in the United States, and more than 400 community colleges and technical schools offer some form of training in geospatial technologies (e.g., see Table A.5 in Appendix A). In contrast, only a handful of U.S. degree or certificate programs have an explicit focus on geospatial analysis. For example, the University of Pennsylvania offers a master's in urban spatial analytics and Duke University offers a geospatial analysis certificate. Various aspects of geospatial analysis are also covered in graduate degree programs in statistics, public health, criminology, archeology, urban planning, ecology, industrial engineering, and other areas. For example, statistics programs with a heavy emphasis on spatial statistics include the University of Minnesota (biostatistics), the University of Washington (environmental and biostatistics), and Duke University (environmental and biostatistics). Advanced courses in spatial optimization are offered in the geography program at the University of California, Santa Barbara, in geography and industrial engineering programs at Arizona State University, and in various programs at Johns Hopkins University and the University of Connecticut.

Training in GIS and geospatial analysis is also delivered through other channels. Professional certificates or degrees are available from traditional or online university programs, both nonprofit and for-profit. Commercial vendors offer professional training or education, typically in the form of online training modules and in-person training sessions. Perhaps the largest and best known industry training is provided by Environmental Systems Research Institute (ESRI), which offers formal technical certification programs that deal with various aspects of GIS and spatial analysis (e.g., desktop, developer, enterprise). Coursework is offered online and in 1- to 4-day instructor-led workshops. After participants pass a test, they are provided with a certificate.

Professional societies (e.g., Association of American Geographers, American Planning Association) sponsor ad hoc training sessions in basic to advanced techniques. These sessions are commonly funded by federal agencies such as the National Science Foundation's Center for Spatially Integrated Social Science, or carried out as part of advanced professional training programs. A number of scholarly conferences include 1- or 2-day short courses or workshops focusing on particular software programs or advanced methods. For example, the GeoStat 2011 conference had a 1-week course on spatial statistics with open-source software.

3

Emerging Areas of Geospatial Intelligence

The National Research Council (NRC, 2010a) report *New Research Directions for the National Geospatial-Intelligence Agency: Workshop Report* identified five emerging subject areas that could potentially improve geospatial intelligence: geospatial intelligence (GEOINT) fusion, crowdsourcing, human geography, visual analytics, and forecasting.[1] Although human geography goes back more than a century, technological and analytical developments have so changed the field that it is treated as an emerging area in this report. Among the emerging areas, there is an emphasis on crosscutting concerns such as three-dimensional and spatiotemporal visualization, as well as linkages between geolocation, social media, crowdsourcing, and spatial analysis. GEOINT fusion covers the linkages, while each of the emerging areas shares the crosscutting concerns.

The five emerging areas are relatively computationally oriented and interdisciplinary, with concepts and skills taught across academic departments. Few are supported by degree programs or academic infrastructure (e.g., professional societies, journals), although these will come as the fields develop. This chapter describes each of the five emerging areas, including its origin, the knowledge and skills that are taught, and the scope of existing education and professional preparation programs.

GEOINT FUSION

GEOINT fusion is concerned with combining geographic information from multiple sources, whether structured or unstructured (e.g., sensor networks, databases, documents), to assess spatial or spatiotemporal phenomena for purposes such as tracking, prediction, or providing a common operational picture. For example, a situation assessment of an ongoing event such as the 2011 Arab Spring may fuse location-aware data from airborne or satellite sensors, social media (e.g., Twitter, blogs), news wires, and reports from observers on the ground. Fusion is important because assessments of a phenomenon from multiple sources of information are likely to be better than those from a single source.

Evolution

Research findings on GEOINT fusion began to be published in the 1980s. Early work provided a classification of use cases (White, 1999) for common tasks such as object refinement (e.g., observation-to-track association, target type and identification), situation assessment (e.g., identification of force structure, communications, and physical context), impact assessment (e.g., consequence prediction, susceptibility and vulnerability assessment), and process refinement (e.g., adaptive search and processing, resource management). These use cases have two dimensions—geographic footprint and temporal extent—as shown in Figure 3.1.

The simplest use case (level 0, subobject) fuses data at the granularity of a single location, such as a pixel in a remote sensing image. For example, a new image

---

[1] Note that these terms differ slightly from those used in NRC (2010a).

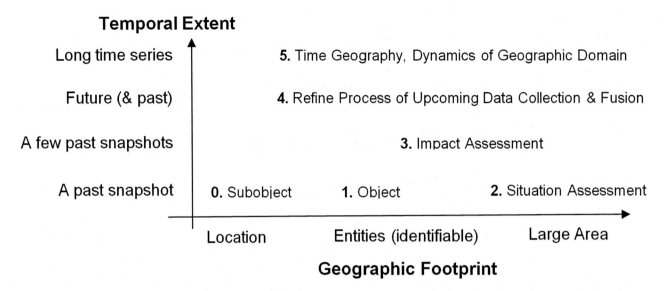

FIGURE 3.1 The complexity and methods used in GEOINT fusion depend both on the size of the geographical area (horizontal axis) and the length of the time period (vertical axis) being covered. In this figure, the classification of use cases is shown by these dimensions.

may be georegistered to a reference map by aligning specific image pixels to corresponding map landmarks. At level 1 (object/entity), information from multiple sensors with overlapping sensing ranges is combined to estimate properties (e.g., location, shape, size, type) of an identifiable entity, such as a vehicle or building. For example, a national air-traffic monitor room may track every aircraft using information collected from local air-traffic controllers. At level 2 (situation assessment), information from all sources is combined to estimate the impact of a recent event or behavior on a geographic area of interest. For example, an emergency manager may fuse weather prediction data sets, plume simulation maps, population density maps, and transportation maps to identify emergency evacuation routes.

The subobject, object, and situation assessment levels are often concerned with a single point in time (snapshot). However, multiple time frames can be considered at any level. At level 3 (impact assessment), a recent image may be compared with an older image to detect major changes in an object or geographic area of interest. For example, the impact of a forest fire may be assessed by comparing remotely sensed images before and after the fire. At level 4 (process refinement), the process of data collection and fusion is refined using what could be considered "control law" that depends on a utility function expressing the dependence of fusion

quality on input quality. For example, fusion may be used to reconfigure the locations and trajectories of sensor platforms (e.g., satellites, aircraft, vehicles) to closely monitor an event (e.g., hurricane) or high-value target in order to improve the quality of fused output estimates of interest.

The late 1990s brought the establishment of the International Society for Information Fusion as well as two journals dedicated to information fusion: *Journal of Advances in Information Fusion* and *Information Fusion*. Conference discussions and publications have refined the use cases in new directions. For example, a long time series of snapshots may enrich traditional fusion with concepts from time-space geography (Hägerstrand, 1967) and the dynamics of geographic domain (Hornsby and Yuan, 2008), leading to a new use case (level 5). At the location (e.g., pixel) geographic footprint, a past time series of measurements can be used to determine a statistical distribution, which, in turn, can be used to evaluate future values for anomalies, regime-change points, and other factors. At an identifiable-entity geographic footprint, a time series of locations produced by multiple sensor measurements for an object can be fused to estimate the object's trajectory, which can be processed further to identify its frequent locations, routes, schedules, and other spatiotemporal patterns (Shekhar et al., 2011).

One may even move beyond events to understand spatiotemporal interactions among event types and underlying processes. For example, a terrorism monitoring and prediction center could use fusion to estimate the parameters of a social-cultural model, which could then be used to assess risks of terrorist attacks at particular locations.

Knowledge and Skills

Fusion draws on many disciplines, including geographic information science, spatial statistics, remote sensing, computer science, electrical engineering, and physics. The concepts are taught at the university level under a variety of topics, such as map conflation (Saalfeld, 1988; Kang, 2009; Longley et al., 2010); spatial statistics (Bivand et al., 2008; Cressie and Winkle, 2011); spatial data mining (Shekhar and Xiong, 2008; Shekhar et al., 2011); data, sensor, or image fusion (Hall and Llinas, 1997; Pohl and Van Genderen, 1998; Hyder et al., 2002; Mitchell, 2010a, b); semantic web (Antoniou and Harmelen, 2004; Allemag and Hendler, 2011); and data, information, or schema integration (Batini et al., 1986; Sheth and Larson, 1990; Lenzerini, 2002; Dyché and Levy, 2006; Halevy et al., 2006). Increasingly this means using an interdisciplinary approach, especially as new data sources (e.g., sensor webs, social network data) are added to existing data sources (e.g., remote sensing). Searching for structure within large volumes of complex, multitheme, and multitemporal data (e.g., big data) also requires interdisciplinary skills, which will become increasingly important as data input sizes continue to grow. "Big data" are often defined by data volumes, variety, and uptake rates that are so large that they challenge the accepted methods of data aggregation, description, visualization, and analysis. Big data present important challenges to GEOINT fusion where current approaches are not scalable. Skills for dealing with these massive data agglomerations may require recruitment of data specialists.

A variety of skills are necessary to handle the workflow to produce GEOINT fusion. For situation awareness, for example, the workflow may include tasks such as identifying relevant sources, georegistering new information (e.g., aerial images), detecting and resolving inconsistencies and uncertainties across sources, characterizing new phenomena from data sources using models, and making cartographic and visualization decisions for presenting the information. Based on common workflows, the necessary skills for fusion include the following:

- Task-relevant source identification. During the 1980s, there were few geospatial intelligence data sources and most of the effort was dedicated to processing. However, advances in sensing, communication, and data management have greatly increased the number of potential sources. As a result, fusion is now leveraging an increasingly diverse array of information sources, including new physical sensors (e.g., videos from unmanned aerial vehicles), social media, and data sets gathered by governments, businesses, and scientists.

- Knowledge of common geospatial intelligence data sources. Data fusion often starts by merging data from multiple sources, which may have different data formats, geographic coordinate systems, geographic resolution, accuracy, and timeliness and are commonly handled by different domain experts. Knowledge of these differences is needed to load data into software systems, to merge data from multiple sources, and to resolve conflicts across data sources.

- Georegistration methods. Fusion often adds new information to a geospatial data set. For example, georegistering information from sources such as aerial imagery, Global Positioning System (GPS) tracks, and cell phones allows information on current locations of friends and foes to be added to a base map. Aerial imagery may be georegistered by identifying several landmarks common to the image and the base map and applying photogrammetric principles. A GPS track may be georegistered to a roadmap in an urban area by identifying the closest roads.

- Deriving new information from sources and managing uncertainty in a complex multisource environment. Some phenomena cannot be fully characterized from observations. Statistical and data-mining methods are used to remove anomalies, identify correlations across data sources, find clusters or groups, and classify or predict specific features using data sources as explanatory features. Evidential reasoning methods such as Bayes' rule or the Dempster-Shafer theory of evidence may be used to estimate the most likely location and shape of a feature from the information available. Optimization techniques from operations research are often required to develop solutions to complex combinatorial optimization problems across all fusion levels. Simulation models may be used to project phenomena, such as trajectories of chemical plumes.

• Geospatial intelligence information presentation. Fusion results are often presented on maps. Preparation of paper maps requires traditional cartographic skills, and preparation of electronic maps requires skills to leverage animation and interaction in context of computer screens, tablets, and cell phones.

• Workflow management. Workflow management systems may be used to specify fusion tasks and their interdependencies as well as to help keep track of progress and facilitate communication among team members. Workflows also enable fusion tasks to be handled within a data collection-analytical context, thus increasing the operational value of the fused data.

Education and Professional Preparation Programs

Although no degree programs are offered in GEOINT fusion, two universities have a research center in fusion: the State University of New York, Buffalo (Center for Multisource Information Fusion) and Pennsylvania State University (Center for Network-Centric Cognition and Information Fusion). In addition, some universities offer courses in various aspects of fusion, largely in computer science departments (e.g., Table A.6 in Appendix A). Semantic web courses are offered by many universities, including Johns Hopkins University, Georgia State University, and Lehigh University. Database interoperability and data integration courses are offered by the University of Southern California and by industry (e.g., Oracle, SAS, Sybase, IBM). Courses in multisensory data fusion are offered by a few universities (e.g., Pennsylvania State University; Arizona State University; Georgia Institute of Technology; University of California, Los Angeles; State University of New York, Buffalo) and by industry (e.g., Objectivity Inc., Applied Technology Institute). In addition, fusion topics are commonly discussed for a few weeks in courses on broader topics at many research universities. For example, geographic information science courses often discuss map conflation, and remote sensing and photogrammetry courses discuss image-to-image and image-to-reference (map) fusion.

In addition, database courses discuss schema integration and data integration; signal processing courses discuss sensor fusion; and statistics, data mining, and spatial computing courses discuss spatial statistics and spatial data mining.

Graduate degrees in related areas (e.g., geographic information science, remote sensing, computer science and electrical engineering) allow a specialization in fusion through research projects and coursework from relevant disciplines. Degrees with a fusion specialization are available from universities with strong education and research presence in geographic information science, remote sensing, spatial statistics, computer science, electrical engineering, and physics. Examples include George Mason University; Georgia Institute of Technology; Ohio State University; Pennsylvania State University; Purdue University; University of California, Santa Barbara; University of Minnesota; and the University of Southern California.

Some professional programs in related broader areas (e.g., geospatial intelligence, geographic information science, security technologies, dynamic network analysis) provide opportunities to specialize in fusion by allowing students to choose a fusion-related capstone project and enroll in fusion-related elective courses. Such training opportunities are available at several universities, including George Mason University; Georgia Institute of Technology; Pennsylvania State University; Redland University; the University of California, Santa Barbara; and the University of Minnesota.

CROWDSOURCING

The term crowdsourcing was introduced by Jeff Howe in a 2006 article in *Wired Magazine* (Howe, 2006) and is defined in the 2011 Merriam-Webster dictionary as "the practice of obtaining needed services, ideas, or content by soliciting contributions from a large group of people and especially from the online community rather than from traditional employees or suppliers."[2] Crowdsourcing is related to participatory sensing, which shares the same principle of collecting data from a set of users working collaboratively (Estrin, 2010). The two terms are often used interchangeably, but the committee prefers the term crowdsourcing,

---

[2] See <www.Merriam-Webster.com>.

which implies not only data collection but also other types of group activities, such as using performing work. Spatial information contributed by crowdsourcing is often referred to as volunteered geographic information. Because such information is collected by volunteers, it comes with challenges of accuracy, credibility, and reliability (Goodchild, 2007; Flanagin and Metzger, 2008). As the use of crowdsourced data grows, issues of data quality, uncertainty, trust, and conflation at the semantic level will increase in importance.

Evolution

Perhaps the earliest example of crowdsourcing was the Longitude Prize, a reward offered by the United Kingdom in 1714 to anyone who could develop a practical method to precisely determine a ship's longitude. Another early example is the 19th Century Oxford English Dictionary, whose editors asked the public to index all words in English and provide example quotations for them (Winchester, 1998). The pace and scale of these volunteer initiatives has increased in recent years with the emergence of the Internet and social networking. Among recent high visibility efforts were the DARPA Network Challenge to collaboratively find marker balloons deployed by DARPA in the United States,[3] and the Netflix Prize to develop algorithms for predicting how well users would like a film, based on their movie preferences.[4] Openstreet Map,[5] an editable map of the world, has been used by numerous companies (e.g., Waze) as their backbone mapping system. Openstreetmap.org had a remarkable success following the Haiti earthquake of January 2010, when volunteers worldwide created a new map from donated imagery in a few days. The crowdsourced map became the most accurate base for relief efforts (Zook et al., 2010).

Today, crowdsourcing plays a major role in creating information-rich maps, collecting geolocalized human activity, and working collaboratively. The convergence of sensing, communication, and computation on single cellular platforms and the ubiquity of the Internet and mobile web have allowed maps to be enriched with a variety of data. Early applications included traffic information collected from smartphones (Figure 3.2;

Cheng, 2009; Hoh et al., 2012), available today from numerous companies (Google, INRIX, NAVTEQ, Waze, BeatTheTraffic.com). The concept was soon extended to enriching maps with other user-generated content, either through location-based services or posting from public records. Examples include maps of crime in Oakland,[6] geolocalized real estate data (e.g., Zillow), photographic geolocalized postings (e.g., Flickr), pedestrian and sports GPS traces (e.g., Nokia[7]), and earthquake information (e.g., Figure 3.3).

The explosion of location-based services has led to the emergence of users sharing personal information (e.g., Facebook), professional information (e.g., LinkedIn), location (e.g., presence in a restaurant, at a landmark location; FourSquare), and social network activities (e.g., placing Facebook activity on maps; Loopt). This new information complements traditional cell tower information, which is already used in operational contexts (e.g., tracking al-Zarqawi by the U.S. military; Perry et al., 2006), by enriching available feeds using attributes disclosed knowingly or not, willingly or not, by the user.

Finally, new concepts of collaborative work are emerging. Wikipedia created a completely crowdsourced encyclopedia on a voluntary basis. It was followed by numerous services provided by volunteers, such as Facebook translation (Hosaka, 2008) and Yahoo! Answers. Amazon's Mechanical Turk[8] enables workers to remotely perform tasks at a distributed and large scale for money. This model represents a new trend in which the crowdsourced workers are active and follow directions. This type of activity has been used successfully for tagging, identification, labeling, parsing, clustering, and recognition.

Knowledge and Skills

Developing the technology for a crowdsourcing system requires knowledge of the problem domain as well as skills in computer programming (including parallel programming), data visualization, database design and management, operating systems, service-oriented architectures, Internet applications, and the

---

[3] See <https://networkchallenge.darpa.mil>.
[4] See <http://www.netflixprize.com/>.
[5] See <http://www.openstreetmap.org/>.

[6] See <http://oakland.crimespotting.org/>.
[7] See the Nokia Sportstracker program at <http://www.sports-tracker.com/>.
[8] See <https://www.mturk.com/mturk/welcome>.

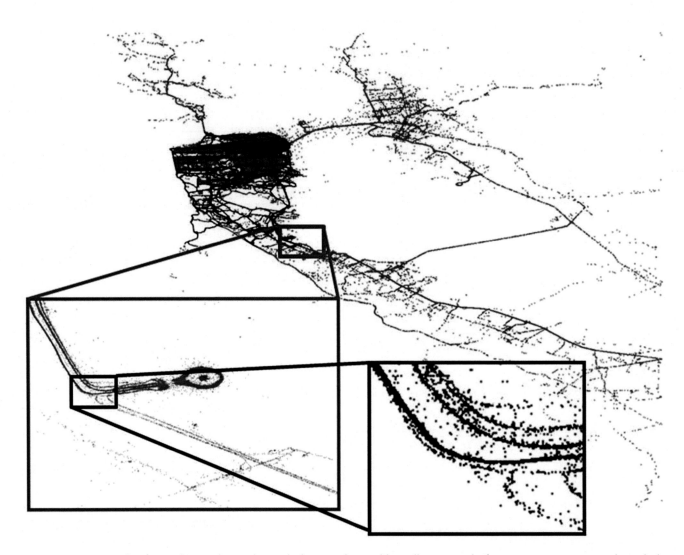

FIGURE 3.2 Example of crowdsourced GPS data, which were obtained by collecting tracks from San Francisco taxis through the Cabspotting program. Each point represents one GPS recording, sampled at an interval of one minute. Three different magnification levels show the detail obtainable from the data. The San Francisco Bay area is shown in red, the approach road to San Francisco International Airport is shown in blue, and the lanes on the Highway 101 intersection by the airport are shown in black. The road map for the Bay Area can be reconstructed from only one day of data. SOURCE: University of California, Berkeley, Mobile Millennium project.

ability to work with various types of data feeds. The technology has been developing rapidly, but a generic set of tools for implementation across applications has yet to emerge.

Building a crowdsourcing system requires the following knowledge and skills:

• Sensing, including hardware knowledge (any type of sensor), device knowledge (using phones or other devices to collect data), and software knowledge (e.g., collecting data from Internet activity).

• Signal processing and filtering, which are needed to remove noise from the data.

• Statistics, machine learning, and large-scale data analytics. Pattern matching, data mining, and statistical inference are needed to extract information from the large volume of data.

• Communications, cellular technology, mobile computing, and human-computer interaction, which are necessary because numerous crowdsourcing systems are based on cellular devices.

• Cloud computing and high-performance computing, which power most crowdsourcing applications.

The knowledge and skills needed to analyze crowdsourced data as well as the crowdsourcing process are

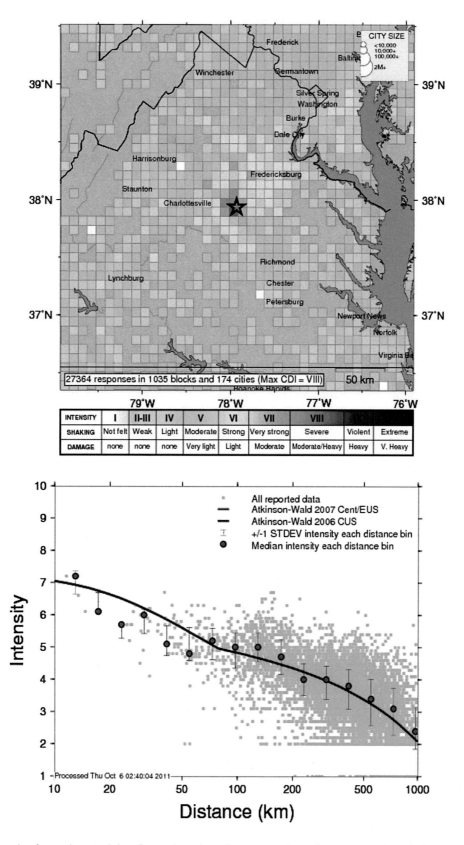

FIGURE 3.3 Example of crowdsourced data for earthquakes. The U.S. Geological Survey's "Did you feel it?" program creates earthquake intensity maps from user responses. The top figure shows the geocoded intensities for the 2011 Virginia earthquake (magnitude 5.8). The bottom figure shows the intensity collected from user input as a distance from the epicenter (dots). The crowdsourced data is compared to model-based predictions (line). SOURCE: U.S. Geological Survey.

markedly different from those required to develop the technology. At a minimum, basic statistical and graphing skills are needed. Additional skills are needed to deal with data tagged with location and temporal information, including econometrics, error estimation, geospatial analytics, geospatial visualization, dynamic analysis, temporal clustering, social network analysis, dynamic network analysis, data mining, and text mining.

Education and Professional Preparation Programs

Crowdsourcing is not an established academic discipline. Students generally gain skills and knowledge in crowdsourcing through special projects carried out as part of a graduate curriculum. Most of the knowledge required for crowdsourcing lies outside traditional geospatial domains, as illustrated by the skills listed above. For this reason, training is distributed among academic departments and programs, including engineering (aerospace, civil, computer science, electrical, environmental, mechanical), statistics, geography, urban planning, and architecture (e.g., Table A.7 in Appendix A). A few multidisciplinary research institutes at universities offer knowledge and skills aligned with training in crowdsourcing, including the following:

• Center for Embedded Networked Sensing at the University of California, Los Angeles, which was one of the first centers to make academic contributions in the field and to offer a doctorate in participatory sensing (Estrin, 2010).
• Computer Science and Artificial Intelligence Laboratory at the Massachusetts Institute of Technology, which has a diverse faculty spanning most of the fields required for crowdsourcing.
• Wireless Information Network Laboratory at Rutgers University, which focuses on privacy and wireless information aspects of crowdsourcing.
• Algorithms Machines People at the University of California, Berkeley, which focuses on building systems that connect people to the cloud to solve hard problems using large data analytics algorithms and massive amounts of crowdsourced and other data.

In most cases, acquiring thorough knowledge

of crowdsourcing requires a doctorate, although implementation skills can be obtained at the master's level. For institutions such as the Massachusetts Institute of Technology, which has a thesis as part of its master's program, or the University of California, Berkeley, which has a project as part of its master's of engineering program, students will gain exposure to the topic through the research or project. In addition, many people involved with crowdsourcing are self-taught and learn by doing. Experts at this level are worldwide and often fall across the age spectrum. Two-year colleges have started to offer curricula to attract these casual practitioners, such as Android phone programming and scripting for web data scraping.

Finally, with the rise of Web 2.0 and the social web, numerous companies have trained engineers in house, enabling them to develop most of the skills needed for crowdsourcing. Several types of companies now have crowdsourcing skill sets, including the following:

• Companies which collect vast amounts of crowdsourced data by the nature of their products, such as Google, Facebook, Twitter, and FourSquare. Each of these companies has divisions or at least groups that focus on the internal development of data analytics tools for crowdsourced data.
• Companies that provide back-end support for systems which rely on crowdsourced data, such as infrastructure systems companies (e.g., IBM, HP) and traffic information companies relying on smartphone data (e.g., NAVTEQ, Waze, INRIX).
• Companies that have developed a business around crowdsourced data analytics, such as SenseNetworks or Sensor Platforms, which were starting up when this report was written.

HUMAN GEOGRAPHY

Human geography concerns the mapping of people, groups, organizations, sentiments and attitudes, norms, belief systems, social activities, and "ways of doing business" over space and time (Figure 3.4). It has been referred to by many names, including cultural geography, human terrain, rich ethnography, cultural mapping, social mapping, sociocultural context, and social

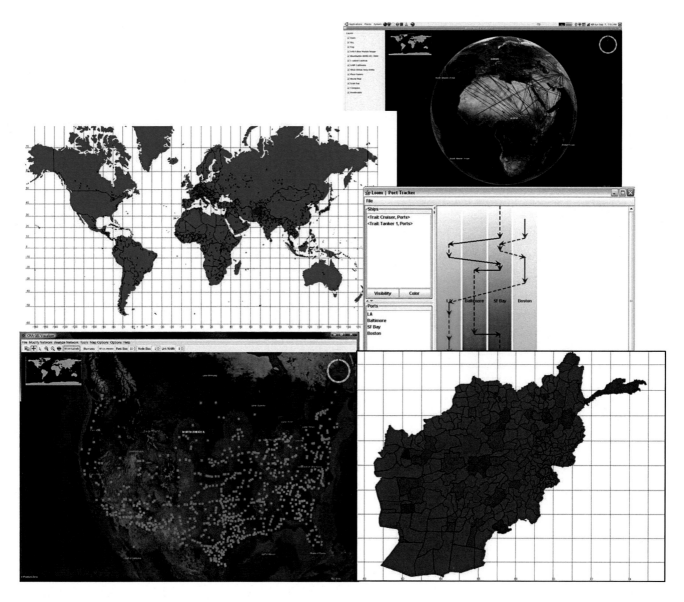

FIGURE 3.4 Different types of visualization used in human geography. (*Top right*) A social network (lines between the locations of participants) superimposed on a NASA Worldwind visualization of the globe. Such images are used to show the relation of social network ties to physical space. (*Middle left*) Map showing the density of a particular activity. Each dot indicates the location where an actor of interest has been seen. The background image is a standard ARCGIS shape file. (*Middle right*) Tracking information for two ships, used to track who or what was where when and to identify common paths. Solid lines show known movement between locations (colored columns) and dashed lines indicate inferred movement or lack of movement. Time is vertical and locations are horizontal. (*Bottom left*) Locations of actors of interest (dots) and secondary information about the spatial density of the betweenness of the nodes (clouds; e.g., Freeman, 1977). Such images are used to identify critical locations. Background image is from NASA Worldwind. (*Bottom right*) Heat map image of Afghanistan using a standard ARCGIS shapefile. Each region is colored by the number of times actors of interest have been in that region. The brighter the red, the higher the level of activity. Such images are used to understand the region of activity and identify points of intervention. SOURCE: All images were produced using ORA.

domain. The use of new technologies and methods, such as network analysis, graph-based statistics, and evolutionary agent-based modeling, distinguishes the emerging area of human geography from its roots as a subfield of geography, sociology, and anthropology.

Evolution

Although human geography has been around for more than a century, the decision to build a human terrain program for the wars in Iraq and Afghanistan led to the rethinking of the role of human cultural knowledge. The human terrain program brought together information technology and a vast array of regional sociocultural information that had been scraped from the web, provided through social media, gathered from other open sources, and collected in the field. The data were analyzed using search and comparison techniques, social network analytics, geographic visualization, and statistics. The aim was to provide up-to-date, accurate information about the general sociocultural environment, current opinion leaders and persons with power, and climate, economic, and political conditions.

Increasingly, sociocultural information, both historical and current, is being placed on maps. New technologies that admit location capture (e.g., modern cell phones) are increasing the amount of location-based data on social and social interaction. Crowdsourcing, Ushahidi-style data captures (e.g., reports submitted by local observers via mobile phone or the Internet), location-based twitters, and so on are providing unprecedented levels of sociocultural information that is at least partially spatially tagged. With new data come new research opportunities and the ability to understand how space constrains and enables social and cultural activity. Illustrative new areas of research include geotemporal social media sampling, location identification from texts, and geonetwork analytics. The next decade will likely see major changes in the quality of sociospatial data presentation and new technologies for capturing, assessing, visualizing, and forecasting social data with a spatiotemporal context.

Knowledge and Skills

Human geography involves four main components:

1. Geo-enabled network analysis—mapping the network of who, what, how, why, and when to locations (e.g., the al-Qaeda social network).
2. Sentiment and technology dispersion—mapping the movement of ideas, activities, technologies, and beliefs as they move from location to location (e.g., the spread of revolution in the Middle East during the Arab Spring).
3. Cultural geography overviews—compendiums of diverse information on current leaders, languages,

foods, habits, religions, etc., which are increasingly taking the form of web-based mashups. Such overviews and the tools for analyzing them formed the basis of human terrain efforts during the Iraq and Afghanistan wars.

4. Sociolinguistic ethnic characterizations— mapping which families, clans, and tribes are where (e.g., the tribal sociolinguistic heredity network).

Each of these areas requires different expertise. Some areas require technical expertise (e.g., programming, scripting) while others require the mastery of advanced conceptual frameworks and approaches (e.g., agent-based modeling, network analysis). These skills are not generally acquired in traditional courses on sensor assessment, cartography, or map interpretation.

An important skill in human geography is text mining: the process of deriving high-quality information from textual sources for analysis. Text, such as news articles, books, twitter feeds, and blogs, contain information about differences in the human condition across locations. Techniques for mining text are reasonably accurate for extracting the names of people, organizations, and locations from English texts. However, challenges remain in interpreting multiple languages, identifying the location of places, and distinguishing between place and person names (e.g., the city of Dorothy Pond, Massachusetts) and place and organization names (e.g., the White House). Both geographical expertise and text-mining expertise are needed to address these problems.

Education and Professional Preparation Programs

A comprehensive human geography program covers five core elements: (1) collection and coding of geomarked human data, (2) geo-enabled text analysis, (3) geo-enabled network analysis and dynamic network analysis, (4) computer simulation of human geography data and forecasts, and (5) geocultural analysis and overviews. Each of these has an associated set of methods and tools that students need to learn, including (1) tools for collecting social media and news data (e.g., TweetTracker, REA); (2) tools for natural language processing, text mining, and sentiment mining (e.g., AutoMap); (3) tools for metanetwork analytics and visualization (e.g., ORA, R); (4) tools for developing agent-based and system dynamic simulations (e.g., MASON, Construct, Dynamo), with particular attention to the diffusion of information and the dispersion of beliefs and activities; and (5) qualitative ethnographic assessment, sociolinguistic characterization, sentiment analysis, text mining, and questionnaires. These elements are rarely taught at the undergraduate or master's level. Most of the education is at the doctorate level (e.g., Table A.8 in Appendix A) or is offered through professional development or specialized training programs such as the Center for Computational Analysis of Social and Organizational Systems (CASOS) Summer Institute. Although many universities cover one or two of these elements in their doctorate programs, only two (Carnegie Mellon University and the University of Arizona) cover all five.

In addition, a number of universities are adding courses in the human-geography area to their doctorate programs. For example, the sociology programs at Cornell and the University of California, Irvine, and the computer science program at the University of Arizona all cover network analysis with courses related to geo-enabled network analysis. The George Mason University Center for Social Complexity and the University of Michigan Center for Complex Systems cover agent-based modeling that takes account of the spatial aspects of human behavior.

Programs that teach social network analysis (Box 3.1) are beginning to cover geo-enabled network analysis. Some programs that teach computer modeling are beginning to teach the programming and data acquisition techniques needed to create and use maps as a way of displaying human behavior. Two-year and community colleges have been among the first academic institutions to teach some of the basic skills needed to use and develop social networking tools and, to some extent, basic tools necessary for network analysis of social data, such as reading GPS signals. These programs are loosely based in media studies and computer science programs and are widespread across the nation.

VISUAL ANALYTICS

Visual analytics is the science of analytic reasoning, facilitated by interactive visual interfaces integrated with computational power and database capacity (Thomas and Cook, 2005). Analytical reasoning is central to the analyst's task of drawing conclusions from a disparate set of evidence and assumptions. The objective of visual analytics is to derive insight from voluminous, changing, vague, and often contradictory geospatial data and other information while avoiding human information overload (van Wijk, 2011). Some examples of information graphics used in visual analytics are shown in Figure 3.5.

Evolution

The growth in the quantities of information that require visual representation and analysis by humans and the increasing complexity of the associated data and analytical problems have given rise to visual analytics as a new scientific discipline (Andrienko et al., 2010). Visual analytics has formalized only recently, with a key publication in 2005 (Thomas and Cook, 2005) and more recently a series of special issues in journals (e.g., Keim et al., 2008; Stapleton et al., 2011).

Visual analytics has origins in cartography, geographic information science, computer vision, information visualization, and scientific visualization. In general, cartography deals with maps and geospatial data, geographic information science deals with spatial relations and spatial query and analysis, scientific visualization deals with data that have a natural physical or geometric structure (e.g., wind flows), and information visualization deals with abstract data structures (e.g., trees, graphs). Choice and reasoning are central to visual analytics.

Research and new directions in visual analytics include creating new information visualization methods, virtual imaging, semantic search, data fusion, dynamic network visualization, and user testing. In particular, methods that focus on how to integrate graphics into the problem-solving process itself has become a key research interest.

Knowledge and Skills

Visual analytics deals with amplifying human cognitive capabilities

- by increasing cognitive capacities and resources, such as memory;

BOX 3.1
Social Network Analysis

There has always been an implicit link between social network analysis and human geography. For example, proximity is a strong basis for individuals forming relations, with most relations weakening with distance. Social network analysis examines the structure of the relations connecting nodes (e.g., people, organizations, topics, events). Many of the earliest studies looked at networks of people connected by relationships such as kinship, mentoring, and works-with. These networks are represented as graphs (e.g., Figure), and matrix algebra or nonparametric network statistics are often used to assess these networks; to identify key nodes, critical dyads, and groups; and to compare and contrast networks (Wasserman and Faust, 1994). Social network analysis is a key methodology in the human geography toolkit.

Evolution. Social network analysis emerged prior to World War II, with early advances in fields such as anthropology, sociology, and communications (Freeman, 2006). The past 10 years have seen a movement to broaden the field of social networks. Changes include the transition from graph-theory-based metrics to a combination of graph-based and statistical measures, the expansion from small networks to very large-scale networks, the increased attention to communication and social media data, and the shift to geotemporal networks. This broader field is often referred to as dynamic network analysis and it is characterized as the study of the structure and evolution of complex sociotechnical systems through the assessment of weighted multimode, multilink, multilevel dynamic networks that are geo-embedded. The field is supported by the quarterly journal *Social Networks*, the online *Journal of Social Structure*, and an increasing number of specialty journals such as *Social Network Analysis and Mining*.

Knowledge and Skills. The study of social networks is integral to fields such as statistics, sociology, organizational science, communication, computer science, and forensic science. However, the ubiquity of networks, the value of graphs as a representation, and the strength of structural thinking has increased the interest in networks in almost every scientific discipline. For example, network analysis has been used in sociology to study social and communications networks (Wasserman and Faust, 1994), in biology to study animal behavior (e.g., Krause et al., 2007), and in geography, civil engineering, ecology, and other disciplines to extend graphs to real or abstract space (Haggett and Chorley, 1969; Urban and Keitt, 2001; Adams et al., 2012). This increased interest has led to a proliferation of theories about how these networks form, evolve, and affect behavior. It has also led to new methods, such as dynamic networks techniques for sets of networks through time, and meta-network metrics for multimode, multilink data. Statistical approaches for assessing dynamics, information loss, and error provide the foundation for social network analysis. Social science approaches are used to study the dynamics within social networks (e.g., reciprocity, social influence, power) and the social, institutional, and historical contexts in which network ties are formed and broken.

Education and Professional Preparation Programs. Classes in social networks are taught in a number of U.S. universities, usually at the doctorate level. However, undergraduate textbooks and courses are starting to appear. Universities with multiple courses in this area include Carnegie Mellon University, University of Kentucky, Northeastern University, Northwestern University, Harvard, Stanford, Indiana University, and the University of California, Irvine. Courses are taught primarily in business and sociology departments, but also in anthropology, communication, management, organizational behavior, organizational theory, strategy, public policy, statistics, information science, and computer science departments. Network analysis in the geometric sense is taught in geography, mathematics, transportation science, computer engineering, and operations research programs.

Continuing education programs provide a primary venue for training in this area. For example, didactic seminars are conducted at the main social networks conference (the International Network for Social Network Analysis) for 2 days prior to the conference. Half-day and full-day training programs are often offered at management science, organization theory, sociology, and anthropology conferences. In addition, there are numerous multiday or week-long training programs, including the CASOS Summer Institute, the Lipari summer school, and the East Carolina University program for marine biologists.

• by facilitating search;
• by enhancing pattern recognition, often by restructuring relations within data;
• by supporting perceptual inference of structures and patterns that are otherwise invisible;
• by improving the ability to monitor large numbers of sensors and events; and

• by providing methods that support exploration and discovery.

Methods in visual analytics are based on principles drawn from cognitive engineering, design, and perceptual psychology (Scholtz et al., 2009). These methods provide a means to build systems for threat analysis, prevention, and response. Visual analytics therefore

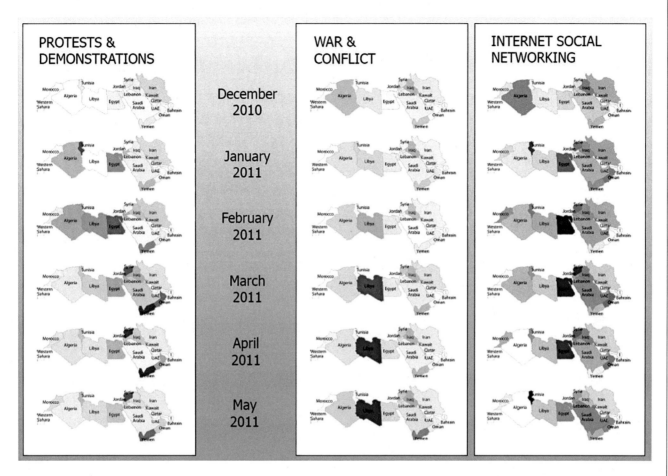

FIGURE Social network analysis is used to show changes in criticality of topics—protests and demonstrations, war and conflict, and Internet and social networking—for the Arab Spring countries. The degree centrality of the three topics (the extent to which a node is connected to other nodes) is based on tags for Lexis-Nexis news articles. The figure shows that the coverage of protests and demonstrations did not spread geographically, and that the change in relevance of the Internet and social networking did not spread in the same way as the revolutions. SOURCE: Courtesy of Jürgen Pfeffer and Kathleen Carley, Carnegie Mellon University.

expands the methods available to analysts but also creates a need for new sets of skills (Ribarsky et al., 2009). Many of these methods are highly dependent on the Internet and on graphics systems and standards.

The suite of skills necessary for research and practice in visual analytics includes an ability to program in scripting and numerical computing languages, an understanding of maps and graphics, the ability to think and

reason spatially, and knowledge of user-centered design principles. For example, programming or scripting skills are needed to develop visualization tools or to extend existing tools, which are commonly targeted to particular applications. Searching for structure within large volumes of complex, multitheme, and multitemporal data (e.g., big data) requires interdisciplinary skills. Severing (2011) noted the importance of moving beyond spe-

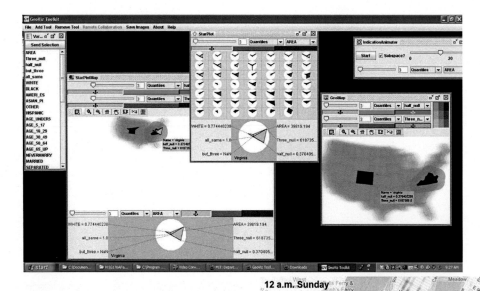

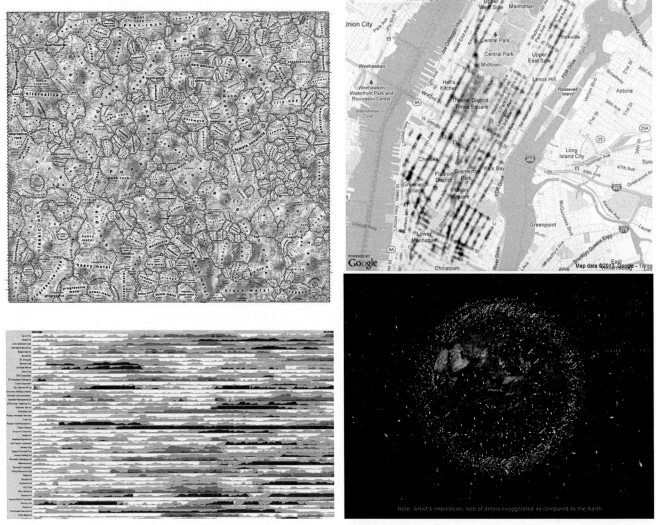

FIGURE 3.5 Examples of some information graphics used in visual analytics. (*Top*) Multimethod display. SOURCE: Screenshot from GeoViz Toolkit developed by Frank Hardisty, GeoVISTA Center. (*Center left*) A semantic landscape of the Last.fm Music Folksonomy using a self-organizing map. SOURCE: Joseph Biberstine, Russell Duhon, Katy Börner, and Elisha Hardy, Indiana University, and André Skupin, San Diego State University, 2010. (*Center right*) Heat map of wireless connections. SOURCE: Sense Networks. (*Bottom left*) Synchronized time-series display. SOURCE: Hannes Reijner, Panopticon Software. (*Bottom right*) Debris objects in low Earth orbit. SOURCE: European Space Agency.

cialization in one field and teaching interdisciplinary flexibility when dealing with big data (in his case for bioinformatics). These skills are rarely available in one person, so teaching and research in visual analytics is commonly carried out by groups of collaborative scholars with different disciplinary backgrounds.

Education and Professional Preparation Programs

Closest to a formal education in visual analytics are interdisciplinary graduate and undergraduate programs that have evolved from communications, visual arts, media studies, geography, computer vision, and human-computer interaction research. For example, at the University of California, Santa Barbara, it is possible to earn a Ph.D. in multimedia arts and technology while doing a considerable amount of coursework in visual analytics. Universities that offer suites of graduate-level classes in visual analytics include the University of North Carolina, Indiana University, the University of Washington, and the Georgia Institute of Technology (Table A.9 in Appendix A). The Georgia Institute of Technology also hosts an online library of materials (e.g., videos, recorded lectures, sample exams) intended for use in higher education in visual analytics.[9]

Methods used in visual analytics are often taught in discipline programs—such as information visualization, cartography, GIS, computer gaming, and computer graphics—although not as a central focus. Many 2-year and community colleges offer basic preparation in visual analytics through media technology, computer programming and scripting, graphic design, imaging and graphics, and human-computer interaction programs.

Research and on-the-job training in visual analytics are also offered by online businesses, gaming companies, and the open-source programming community. Visual analytics research has bases at both the Pacific Northwest National Laboratory and at the Oak Ridge National Laboratory. Private companies involved in visual analytics include Northrop Grumman and Oculus, Inc., a Toronto-based company working on the visual display of time-space tracks. The primary avenue for discussing visual analytics is national conferences, most based in the United States, such as

InfoViz, Where2.0, and the Institute of Electrical and Electronics Engineers Symposium on Visual Analytics Science and Technology.

FORECASTING

Forecasting is a technique that uses observations, knowledge about the processes involved, and analytical skills to anticipate outcomes, trends, or future behaviors. Forecasts are related to predictions and anticipatory intelligence. In general, forecasts attempt to estimate a magnitude or value at a specific time (such as 3-day forecast of temperature), whereas predictions estimate what may happen and the odds of it happening (such as predicting what fraction of people will develop skin cancer). Anticipatory intelligence combines computational methods (e.g., agent-based modeling, system dynamics, Bayes network models) with role playing and applications of game theory to generate integrated time-based simulations.

In the geospatial domain, forecasting needs to address what, where, when, and how events will unfold and how processes will evolve in space and time. Geospatial events and processes are a result of interactions among the natural and built environments as well as social and cultural systems across global, regional, and local scales.

Evolution

The ability to forecast future behavior is central to many scientific disciplines. Among the first disciplines to embrace quantitative methods for forecasting were meteorology and economics. Weather forecasts were made from data, charts, and maps until the late 1950s, when empirical methods began to be replaced by numerical weather forecasting (Lutgens and Tarbuck, 1986). Similarly, economic forecasts transitioned from methods using stationary and deterministic assumptions in the late 1960s (Khachaturov, 1971) to probabilistic or stochastic methods, then to complex simulations of dynamic, adaptive economic systems in 1990s and 2000s (Clements and Hendry, 1999; Gasparikova, 2007).

Recent advances in computational methods, econometrics, simulation, system dynamics, agent-based modeling, and game theory have allowed forecasters

---

[9] See <http://vadl.cc.gatech.edu>.

to generate a range of possibilities to support decision making or scenario-based planning. The International Institute of Forecasters (IIF) was founded in 1981 to promote forecasting through multidisciplinary research, professional development, bridging theory and practice, and international collaboration among decision makers, forecasters, and researchers. A majority of its members are from the economics, business, and statistics communities. IIF publishes two journals: the *International Journal of Forecasting* (a peer-reviewed academic journal started in 1985), and *Foresight: The International Journal of Applied Forecasting* (a journal for practitioners, started in 2005).

Advances in sensor technologies and the increasing availability and timeliness of information have opened new opportunities for forecasting. Forecasts are now being made in areas ranging from ecology (Luo et al., 2011) to technology (NRC, 2010b) to sports (Yiannakis et al., 2006). The concept of nowcasting—forecasts of local events in near-real time—has emerged for both physical and socioeconomic systems. For ex-

ample, nowcasting systems to project the development and dissipation of convective storms 2 hours ahead were tested during the 2008 Beijing Olympics (Wilson et al., 2010). Nowcasting is considerably more challenging than forecasting. It is one thing to forecast population growth of a city over the next year; it is quite another to nowcast the population distribution downtown for emergency evacuation. Nowcasting demands rapid assimilation of massive amounts of data from multiple sources into model runs; scientific understanding of event evolution, the environment, and their interactions; and the ability to deal with measurement errors, incomplete data, or uncertain information in real time. Moreover, research shows that both computational models and human judgment are required to optimize the nowcast (Monti, 2010).

Geospatial intelligence forecasting can play a key role in informing a variety of decisions for military or security operations. Examples include determining optimal clothing based on weather forecasts (Morabito et al., 2011), strategic planning based on forecasts of politi-

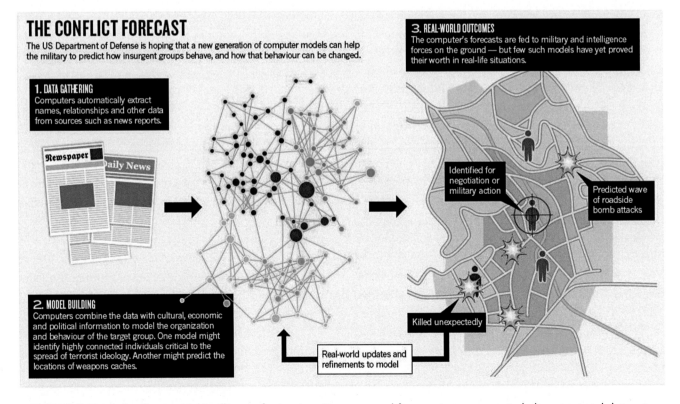

FIGURE 3.6 Example of geospatial intelligence forecasting. Data extracted from various sources, including structured data sets or unstructured texts, provide information about people, activities, and events. The information is analyzed using computer models to reveal the potential connections among people, activities, and/or events and to project possible future events. SOURCE: Reprinted by permission from Macmillan Publishers Ltd on behalf of Cancer Research UK: Web of War, Weinberger (2011).

cal instability (Goldstone et al., 2010) or other events that may threaten liberal democracies (Anderson, 2010), and anticipating social or political change through cyber-empowered political movements, social disruptions, or cultural conflicts (Bothos et al., 2010; Paris et al., 2010; Weinberger, 2011; Figure 3.6). However, rigorous methods for forecasting social patterns and social changes have not yet been fully developed.

Knowledge and Skills

Robust forecasting methods build on a solid understanding of the composition and structure of a system and the embedded interactions among system components and between the system and its environment (Boretos, 2011). Geospatial forecasting requires both deep domain knowledge and advanced skills in spatiotemporal analysis, modeling, and synthesis. Examples include regression statistics, spatial and temporal interpolation techniques, space-time prisms and trajectory models, cellular automata and agent-based modeling, artificial neural networks, evolutionary and genetic algorithms, computer simulation and ensemble techniques, and scenario-based planning that anticipates multiple possibilities.

Forecasts in the context of geospatial intelligence need to integrate both geospatial processes and domain processes to reveal patterns, relationships, and mechanisms that drive state changes. For example, activity-based intelligence—the predictive analysis of the activity and transactions associated with an entity, population, or area of interest—depends on an understanding of environmental, social, and cultural factors; individual space-time behaviors; and the spatiosocial processes that move and regulate activities of groups and the society.

New methods and analytical tools emerging from the computational social sciences are changing the education and skills needed for geospatial intelligence forecasting. For example, new approaches are being developed to address the validation and calibration challenges of agent-based and other complex systems models. Tools such as the Integrated Crisis Early Warning System have been developed to predict political events such as insurgency, civil war, coups, or invasion. The increase in volunteered geographic information and geotagged images or communica-

tions brings the field a step closer to short-term and near-real-time forecasts of event progression, such as the spread of wildfire or disease, or of social dynamics, such as perception or activities planning.

Education and Professional Preparation Programs

No university programs offer degrees in forecasting, and many science-based or business-based curricula emphasize modeling instead of forecasting. Courses in advanced methods for spatial and domain-specific processes are taught at senior undergraduate or graduate levels in a wide range of disciplines, including statistics, computer science, information science, electrical engineering, civil engineering, meteorology, geography, economics, ecology, criminology, epidemiology, and urban and regional planning. Geospatial forecasting requires an integrative treatment of spatial and temporal data and is still considered an advanced, specialized area of research. The few advanced spatial modeling courses available are commonly tailored to the faculty's research interest, rather than providing a comprehensive coverage of analytical and modeling techniques. Examples of universities with strong programs in agent-based modeling include Carnegie Mellon University, George Mason University, and the University of Michigan (see Table A.10 in Appendix A). The Massachusetts Institute of Technology has a strong program in system dynamics.

Time-series analysis is the foundation for forecasting, and relevant courses are commonly taught in meteorology, geography, geology, ecology, economics, political science, and other departments that emphasize modeling and projections. Students learn how to detect temporal trends and to project them into the future using techniques such as harmonic analysis, wavelet analysis, and historical event modeling. Examples of programs that offer courses in these areas include the University of Oklahoma and the University of Washington (meteorology); the University of California, Santa Barbara, and the State University of New York at Buffalo (geography); and Harvard University and Princeton University (economics and political science; see Table A.10 in Appendix A).

Space presents another important dimension of forecasts. In human geography, spatial diffusion theory, central place theory, and time geography offer both conceptual and mathematical bases for spatial prediction, such as spatial interpolation, spatial gravity modeling, spatial regression, and spatial optimization. These traditional analog and mathematical modeling techniques are commonly taught in geography, geology, epidemiology, criminology, civil engineering, transportation science, urban and regional planning, and landscape architecture departments. A few universities offer advanced geocomputational methods for spatial prediction, such as Monte Carlo simulation, Markov chain modeling, cellular automata, agent-based modeling, geographically weighted regression, spatial self-organizing maps, spatial trajectory modeling, spatial niche modeling, spatial Bayesian statistics, and spatial econometrics. Example universities offering courses in the spatial aspects of forecasting include Arizona State University; Clark University; the University of Texas, Dallas; San Diego State University; the University of Utah; the University of Maryland; and Ohio State University.

Some community colleges or technology centers (e.g., GeoTech Center) offer basic statistics courses or computer modeling tools (such as STELLA), which can provide foundation training for beginners. Opportunities for professional training in forecasting are limited. Workshops or summer schools, such as those offered by the Spatial Perspective to Advance Curricular Education program,[10] the Center for Spatially Integrated Social Science,[11] and the University of Michigan, are perhaps the main form of training for advanced space-time methods or geocomputational techniques. Many of these workshops cover only the fundamentals. For economics and business, the IIF frequently offers training workshops for practitioners at their conferences.

---

[10] See <http://www.csiss.org/SPACE/>.
[11] See <http://www.csiss.org/>.

4

Availability of Experts

Applicants for National Geospatial-Intelligence Agency (NGA) positions must be U.S. citizens and have a relevant bachelor's degree, experience, or both (Box 4.1). Many geospatial intelligence disciplines needed for these positions are small and some are shrinking. Thus, a key question for agency managers is how many individuals have education and/or experience in the core and emerging areas now and over the next 20 years (Task 1). This chapter assesses the supply of two sources of expertise in the core and emerging areas: (1) new graduates entering the workforce with a relevant degree, and (2) individuals already working in occupations outside NGA that require relevant knowledge or skills. It also examines factors that reduce the availability of this expertise to NGA.

CURRENT AVAILABILITY OF EXPERTS

To assess the current availability of experts in the core and emerging areas, the committee analyzed government statistics on the number of individuals graduating with a relevant degree from a U.S. college or university and the number of experienced individuals employed in occupations that require relevant knowledge or skills. For example, one source of NGA employees is former military officers, who have received a substantial amount of on-the-job training in the field of service. Data on new graduates are available from the Department of Education, which tracks the number of degrees conferred by level and field of study. Data on experienced individuals working in related fields are available from the Bureau of Labor Statistics, which

collects and tracks employment statistics for more than 800 occupations.

Supply of New Graduates

The Department of Education's National Center for Education Statistics gathers information from U.S. colleges and universities on the number of degrees conferred by degree level and field of study through its Integrated Postsecondary Education Data System. Educational institutions use a set of instructional programs defined in the Classification of Instructional Programs to report degree data. The Classification of Instructional Programs includes more than a thousand programs.[1] Cartography is the only core or emerging area tracked in this classification directly, but some of the other areas are mentioned in the code descriptions. For example, photogrammetry appears in the descriptions of three codes; remote sensing appears in six codes; and geodesy appears in four codes. Because the core and emerging areas do not exactly correspond to the instructional program codes, the committee chose relevant fields of study to track by matching the descriptions of the instructional programs to the skills, degrees, or coursework identified for the core and emerging areas (Chapters 2 and 3). The committee used expert judgment to rank each instructional program as highly relevant, possibly relevant, or not relevant to each area.

---

[1] See <http://nces.ed.gov/pubs2002/cip2000/ciplist.asp> for a list of all programs. The committee used the 2000 version of the Classification of Instructional Programs because data were not available for the 2010 version when the analysis was carried out.

BOX 4.1
NGA Education Requirements

Most geospatial analysis positions at NGA require a bachelor's degree, 6 years of experience, or a comparable combination of education and experience (Table B.1, Appendix B). A master's or doctorate degree is preferred for principal scientists. In NGA's current workforce, more than 80 percent of scientists and analysts have a bachelor's degree and 30 percent have a master's degree. Relatively few have a doctorate degree or less than a bachelor's degree (e.g., an associate's degree).

The fields of study specified in the education requirements for NGA scientist and analyst positions are diverse (Table B.1, Appendix B). A few occupations have relatively specific requirements. For example, analysts specializing in geospatial analysis must have a related degree or a certificate in Geographic Information Systems from an accredited university. However, many positions allow a wide range of degree topics. For example, applicants for imagery intelligence positions may hold a bachelor's degree in engineering, foreign area studies, geography, history, imagery science, international affairs, military science, physical science, political science, remote sensing, or a related discipline. The 25 most common degree topics specified in science and analysis positions at NGA are shown graphically in the figure below. The most highly sought degree topics—physical science, engineering, mathematics, and geography—are broad areas that encompass several fields of study, suggesting that NGA is flexible on the field of study for its science and analysis positions.

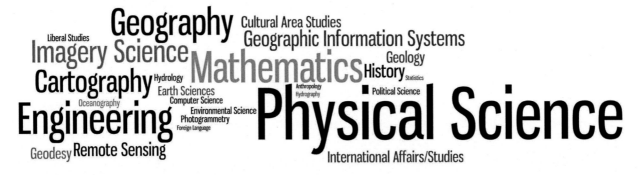

FIGURE The 25 most common degree topics specified in NGA job descriptions for scientists and analysts (Table B.1, Appendix B), which focus on five core areas: geodesy and geophysics, photogrammetry, remote sensing, cartography, and GIS and geospatial analysis. In word clouds such as this, the most common topics are portrayed by the largest lettering. An arbitrary color scheme is used to make it easier to distinguish the various phrases. SOURCE: Generated using <http://www.wordle.net>.

A total of 164 instructional programs were judged to be highly relevant or possibly relevant to the core and emerging areas. The definitions of these programs are given in Table C.1 of Appendix C, and the assignment of relevance to each core and emerging area is given in Table C.2.

Figure 4.1 shows the number of instructional programs that potentially provide knowledge and skills that are relevant to each of the core and emerging areas. Areas that are highly interdisciplinary (e.g., human geography, forecasting) or that are taught in several different university departments (e.g., remote sensing) have the largest number of highly relevant instructional programs (21–57). The area with the lowest number of relevant instructional programs is photogrammetry, which has only 1 highly relevant instructional program

(surveying engineering) and 12 possibly relevant programs. The highly relevant instructional programs that were identified most often across the core and emerging areas were surveying engineering, mathematical statistics and probability, and cartography.

The large number of highly relevant or possibly relevant instructional programs (164) yields a correspondingly large number of graduates. For human geography, for example, the committee deemed 54 instructional programs as highly relevant and 28 instructional programs as possibly relevant. In 2009, more than 150,000 degrees were conferred at all levels for the highly relevant instructional programs and nearly 100,000 degrees were conferred for the possibly relevant instructional programs (Table C.3 in Appendix C). These numbers are clearly overestimates of the recruitment pool, given

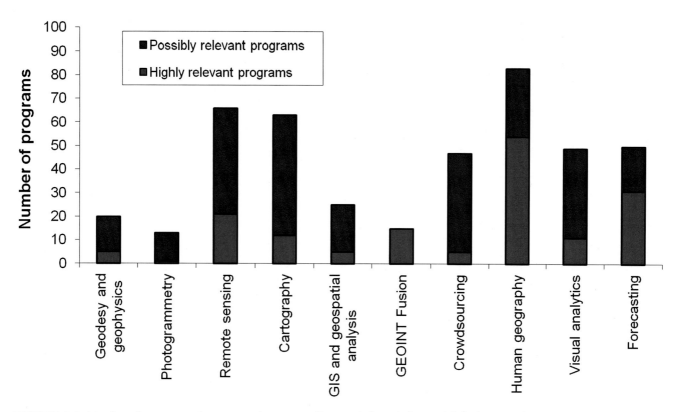

FIGURE 4.1 Number of instructional programs that potentially provide knowledge and skills that are relevant to the core and emerging areas. SOURCE: Data from the National Center for Education Statistics' Classification of Instructional Programs, <http://nces.ed.gov/pubs2002/cip2000/ciplist.asp>.

that human geography, as defined in this report, is an emerging area.

A more realistic "upper bound" on the number of graduates was determined by focusing on the 109 instructional programs considered by the committee to be highly relevant to the core and emerging areas. Figure 4.2 shows the number of bachelor's, master's, and doctorate degrees conferred in 2009 in highly relevant fields of study (see also Table C.4 in Appendix C). For each of the core and emerging areas, a few instructional programs produce more than 50 percent of graduates (Table C.6):

- geodesy and geophysics: aerospace, aeronautical and astronautical engineering;
- photogrammetry: surveying engineering;
- remote sensing: mathematics, general;
- cartography: geography; graphic design;
- GIS and geospatial analysis: geography;
- GEOINT fusion: information science/studies; information technology; environmental studies; environmental science;

- crowdsourcing: information technology; statistics, general;
- human geography: political science and government, general; history, general; sociology;
- visual analytics: information science/studies; graphic design; and
- forecasting: political science and government, general; sociology.

The instructional programs that produce the bulk of graduates do not always match the programs that provide the bulk of skills needed for a position in a core or emerging area (e.g., see Tables A.1–A.10, Appendix A). The mismatch is greatest in remote sensing, geodesy and geophysics, human geography, and forecasting.

Figure 4.2 shows that more than three-quarters of the degrees were at the bachelor's level and about 18 percent were at the master's level. The mix of degrees conferred varied among the core and emerging areas, with a larger fraction of bachelor's degrees in fields highly relevant to cartography, human geography, and forecasting; a larger fraction of master's degrees in

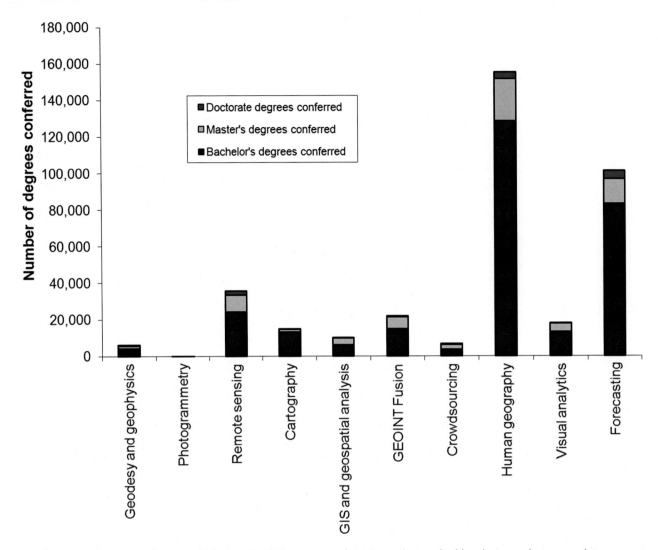

FIGURE 4.2 Number of degrees conferred in 2009 in instructional programs that are highly relevant to the core and emerging areas. SOURCE: Data from the U.S. Department of Education, National Center for Education Statistics' Integrated Postsecondary Education Data System Completions Survey. Accessed via WebCASPAR.

fields highly relevant to crowdsourcing and GIS and geospatial analysis; and a larger fraction of doctorate degrees in fields highly relevant to geodesy and geophysics and remote sensing.

Figure 4.3 shows trends in the number of graduates for all levels in fields of study that are highly relevant to the core and emerging areas for the 2000–2009 period (see also Table C.4, Appendix C). The figures show that the total number of degrees conferred grew over the 2000–2009 period. Annual growth rates for that period were 3.5 percent for bachelor's degrees, 4.5 percent for master's degrees, and 2.1 percent for doctorate degrees.[2]

Growth rates in the number of degrees conferred in highly relevant fields of study vary considerably by area (Figure 4.4; Table C.5, Appendix C). The areas with the highest growth in the number of relevant degrees (annual growth rates greater than 4 percent) from 2000 to 2009 were geodesy and geophysics, cartographic science, crowdsourcing, and visual analytics. Some of the fields of study driving the increase include aerospace, aeronautical, and astronautical engineering (geodesy and geophysics); information technology

---

[2] Annual growth rates were calculated using the standard compound annual growth rate formula: (ending value ÷ beginning value)$^{1/N}$ − 1.

In this case, the ending value is the number of degrees conferred in 2009, the beginning value is the number of degrees conferred in 2000, and N is the number of years that have elapsed between the beginning and ending values (9 years).

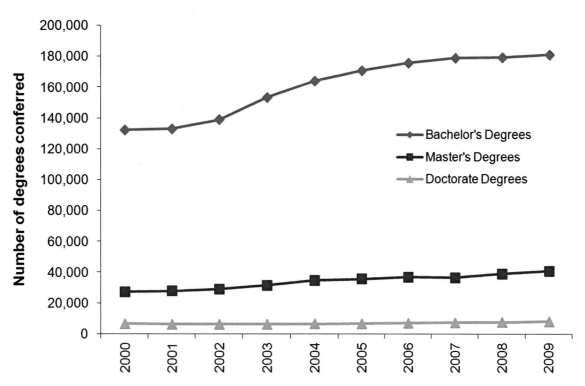

FIGURE 4.3 Number of degrees conferred by year for the fields of study that are highly relevant to the core and emerging areas. SOURCE: Data from the U.S. Department of Education, National Center for Education Statistics' Integrated Postsecondary Education Data System Completions Survey. Accessed via WebCASPAR.

(crowdsourcing); and graphic design, and animation, interactive technology, video graphics, and special effects (cartographic science and visual analytics). The sharp increase in 2003 or 2004 for crowdsourcing, cartography, and visual analytics (Figure 4.4 top) reflects the introduction of new instructional program codes (Table C.6, Appendix C). The 2008 decline in crowdsourcing reflects a decreasing number of degrees conferred in information technology. The decline for visual analytics between 2003 and 2007 reflects a decrease in degrees conferred in information science/ studies, although recent increases in degrees conferred in animation, interactive technology, video graphics, and special effects have led to a recent uptick. No clear trend is apparent in photogrammetry, possibly because only one field of study was considered highly relevant and numbers of graduates in that field are small and were not reported until 2004.

The degree data compiled by the Department of Education are not ideal for estimating the supply of new graduates in the core and emerging areas discussed in this report. One shortcoming is that the

programs included in the Classification of Instructional Programs may not perfectly match university programs. Each university uses discretion in matching the degrees it confers to the programs included in the classification, which could result in inconsistencies across universities. Moreover, instructional program codes evolve over time. For example, nearly half of the instructional programs that are highly relevant to the core and emerging areas were introduced in the classification used in this study, one was discontinued, and several were renamed. In some of the new fields, no degrees were reported, suggesting that it takes time for universities to adopt new classifications.

The most important shortcoming in the Department of Education data is that only one instructional program (cartography) directly matches an area analyzed in this report. For the other core and emerging areas, several instructional programs potentially offer some relevant knowledge and skills. Adding up the number of graduates from all relevant instructional programs yields an "upper bound" on the number of experts in the core and emerging areas. To estimate

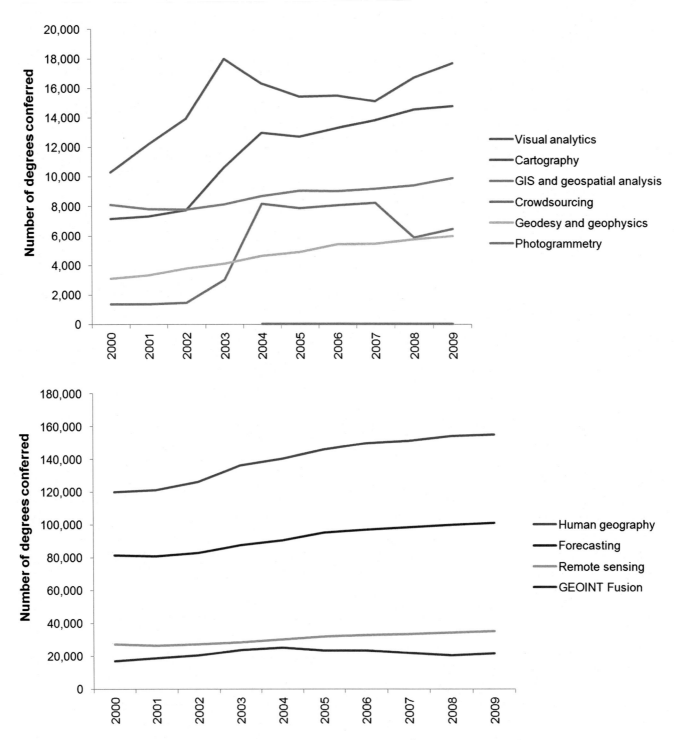

FIGURE 4.4 Total number of degrees conferred for 2000–2009 by year for fields of study that are highly relevant to the core and emerging areas. Note the change in scale between the two figures. (*Top*) Ten-year trends for photogrammetry, geodesy and geophysics, crowdsourcing, GIS and geospatial analysis, cartography, and visual analytics. (*Bottom*) Ten-year trends for GEOINT fusion, remote sensing, forecasting, and human geography. SOURCE: Data from the U.S. Department of Education, National Center for Education Statistics' Integrated Postsecondary Education Data System Completions Survey. Accessed via WebCASPAR.

the number of graduates with the desired mix of skills and knowledge, the committee took into account the number of graduates from the prevailing instructional program (if any) and the number of universities offering programs in a core or emerging area (see Tables A.1–A.10, Appendix A). For remote sensing, which has no prevailing instructional program, the committee made its estimate based on the number of universities and the membership of the primary professional society for remote sensing and related geospatial analysis—the American Society for Photogrammetry and Remote Sensing (ASPRS). The committee's estimates, which are based on expert judgment, are given in Table 4.1.

Accounting for the number of university programs in the core and emerging areas and the instructional programs most commonly associated with these programs lowers the estimated number of graduates in the core and emerging areas (Table 4.1). The difference between the total number of graduates from highly relevant instructional programs and the number of graduates with knowledge in a core or emerging area is smallest for photogrammetry and for GIS and geospatial analysis, each of which is dominated by one instructional program that provides the bulk of training necessary for the core area (i.e., geography for GIS and geospatial analysis [Table A.4]). For core or emerging areas with a small number of graduates in a closely related instructional program (e.g., geodesy and geophysics, cartography) and/or a small number

TABLE 4.1 Estimated Annual Number of Graduates at All Levels with Knowledge in a Core or Emerging Area

Number of Graduates in Relevant Fields of Study[a]	Number of Universities with Training in a Core or Emerging Area[b]	Number of Graduates with Knowledge in a Core or Emerging Area
Geodesy and geophysics • 5,979 total graduates • 213 geophysics and seismology graduates • 28 surveying engineering graduates	60 universities for geophysics 20 universities for geodesy	hundreds
Photogrammetry 28 total graduates, all in surveying engineering	15 universities	few tens
Remote sensing 35,427 total graduates	63 universities	hundreds to thousands[c]
Cartographic science • 14,779 total graduates • 165 cartography graduates	35 universities	hundreds
GIS and geospatial analysis • 9,917 total graduates • 5,615 geography graduates	189 universities	thousands
GEOINT fusion 21,656 total graduates	12 universities	tens to hundreds
Crowdsourcing 6,469 total graduates	10 universities	tens to hundreds
Human geography 155,016 total graduates	10 universities	tens to hundreds
Visual analytics 17,678 total graduates	15 universities	tens to hundreds
Forecasting 101,121 total graduates	100 universities	hundreds to thousands

[a] See Table C.6, Appendix C.
[b] See Tables A.1–A.10, Appendix A.
[c] Based on the number of universities and the membership of ASPRS (7,000).

of university programs (e.g., all emerging areas), the difference in estimates can be several orders of magnitude (Table 4.1). The refined estimate suggests that the number of graduates with expertise in the core and emerging areas ranges from tens (e.g., photogrammetry, crowdsourcing) to thousands (e.g., GIS and geospatial analysis).

Supply of Experienced Individuals

The Bureau of Labor Statistics' Occupational Employment Statistics program estimates the number of jobs and wages for more than 800 occupations. The number of jobs in a specific occupation is similar to, but not the same as, the number of people employed in that occupation. For example, a person may have more than one job. Thus, these job estimates are not direct estimates of the number of people available in a given occupation. The job and wage estimates are based on a survey of more than a million business establishments in the United States, the District of Columbia, Guam, Puerto Rico, and the U.S. Virgin Islands over a 3-year period. The data do not reflect the self-employed, owners and partners in unincorporated firms, household workers, and unpaid family workers.

The Occupational Employment Statistics program classifies occupations using the Standard Occupational Classification system. The codes and descriptions for the 36 occupations chosen by the committee as most relevant to NGA are given in Table D.1 in Appendix D. Some of the occupations are closely related to the core and emerging areas, such as cartographers and photogrammetrists; surveying and mapping technicians; geographers; and geoscientists, except hydrologists and geographers. The individuals working in these occupations likely have knowledge and skills that would be useful to NGA. However, most occupations are more broadly defined than the core and emerging areas (e.g., computer programmers, computer systems analysts, electrical engineers) and likely include workers with skills and knowledge that are less relevant to NGA.

Table D.2 in Appendix D lists the number of jobs and wages by sector as of May 2010 for NGA-relevant occupations. The data suggest that there are more than 2.7 million jobs in occupations that are potentially relevant to NGA. A relatively modest percentage of these jobs are in the federal sector (8 percent), with the vast

majority (77 percent) in the private sector. The federal sector employs more than 50 percent of the nation's forest and conservation technicians, geographers, and political scientists. The bulk of jobs in the private sector are in computer occupations, with more than half in three occupations: software developers, applications; computer systems analysts; and computer programmers. Occupations that are likely to be particularly relevant to NGA—cartographers and photogrammetrists; surveying and mapping technicians; geographers; and geoscientists, except hydrologists and geographers—represent less than 4 percent of the jobs in NGA-relevant occupations.

For a majority of occupations, the mean annual salary is higher in the federal sector than in the private sector (Table D.2, Appendix D). These data are consistent with a Congressional Budget Office report (CBO, 2012), which found that average wages and benefits are higher for federal workers with bachelor's and master's degrees than for private-sector workers. As noted by the Bureau of Labor Statistics, however, these salary differences may be due to factors such as the level of work performed, age and experience, cost of living, establishment size, work schedules, and unionization.[3] Given these caveats, the occupations with the most notable differences in salary are astronomers and historians, which have mean annual salaries that are more than 50 percent higher in the federal sector than in the private sector. Pay for mathematical technician occupations is notably lower in the federal sector than in the private sector.

As noted above, former members of the military are a source of employees for NGA. Statistics from the American Community Survey of the U.S. Bureau of Census can be used to estimate the fraction of people in NGA-relevant occupations who are currently serving or have served in the military. The data for 2010 show that current and former military make up approximately 11 percent of employees in NGA-relevant occupations, including 11 percent of surveyors, cartographers, and photogrammetrists; 16 percent of surveying and mapping technicians; 16 percent of miscellaneous social scientists, including survey researchers

---

[3] Bureau of Labor Statistics, U.S. Department of Labor, Occupational Employment Statistics, Frequently Asked Questions, <http://www.bls.gov/oes/oes_ques.htm>.

and sociologists; and 40 percent of atmospheric and space scientists (Table D.3).

Individuals who are unemployed in NGA-relevant occupations discussed above may also be a viable source of expertise. Information on unemployment by occupation is available from the Bureau of Labor Statistics' Current Population Survey. Coverage of the survey is limited to the civilian noninstitutional population aged 16 years and older and involves a monthly survey of 60,000 households. The occupation data collected are consistent with the 2000 Standard Occupation Code system but are presented at a higher level of aggregation in some cases.

Table D.4 in Appendix D shows average annual unemployment rates[4] for salary and wage workers in NGA-relevant occupations over the 2006–2010 period. The average annual unemployment rate for management, professional, and related occupations as a whole (5 percent for 2010) was included as a benchmark. Two occupations with notably high total unemployment rates for 2010 were surveying and mapping technicians (15 percent) and artists and related workers (14 percent). These two occupations have experienced relatively high unemployment rates since 2006, suggesting that the high rates in 2010 were not due solely to the recent recession.

Occupations with relatively low 2010 unemployment rates include statisticians (<1 percent); urban and regional planners (1 percent); operations research analysts (2 percent); environmental scientists and geoscientists (2 percent); librarians (3 percent); and physical scientists, all other (3 percent). The 2010 unemployment rate for surveyors, cartographers, and photogrammetrists (2 percent) was lower than the management, professional, and related occupations benchmark in 2010, but higher in most other years.

The labor data are subject to some of the same shortcomings as the degree data. In particular, only a few occupations match the core areas and none match the emerging areas. Thus, the 2.7 million jobs in NGA-relevant occupations provide an "upper bound" on the number of experienced workers with some knowledge

or skills needed for the core and emerging areas. The actual number is likely considerably lower. A possible "lower bound" is the number of jobs in the four most closely related occupations: cartographers and photogrammetrists (11,670); surveying and mapping technicians (53,870); geographers (1,300); and geoscientists, except hydrologists and geographers (30,830).

REDUCTIONS IN THE TALENT POOL

The talent pool available to NGA is smaller than the estimates presented above because only U.S. citizens able to obtain a security clearance are eligible for hire. In addition, competition from other organizations, which may offer higher salaries or a better work environment, may reduce the number of highly qualified applicants. For example, new graduates are accustomed to staying connected with their peers, downloading applications, and using any software they wish to carry out tasks on computer platforms and mobile devices. Organizations that do not offer such a flexible, high-tech, connected environment, such as government agencies, may not attract the most technically savvy and analytically capable individuals. Moreover, some individuals will not work for the government or for an intelligence agency. It is difficult to quantify reductions in the labor pool associated with the NGA work environment, but data are available to assess reductions associated with the U.S. citizenship requirement, as discussed below.

U.S. Citizenship

The citizenship of new graduates can be inferred from data gathered by the National Center for Education Statistics. Although these data do not distinguish between U.S. citizens and permanent residents, other data sources suggest that the vast majority of degrees conferred to U.S. citizens and permanent residents are conferred to U.S. citizens. In particular, an analysis of the most recent cohort in the Baccalaureate and Beyond Longitudinal Study,[5] representing bachelor's degree recipients during the 2007–2008 academic year, shows that more than 96 percent of degree recipients were U.S. citizens and 3 percent were permanent resi-

---

[4] The unemployment rate is the number of people who are unemployed (i.e., people without jobs who are looking for work) divided by the number of people in the labor force (i.e., employed people plus unemployed people). See Table D.4 for details on how the unemployment rate is calculated.

[5] See <http://nces.ed.gov/surveys/b&b/>.

dents. An analysis of the most recent National Survey of Recent College Graduates,[6] representing individuals who earned a bachelor's or master's degree in a science, engineering, or health field between July 1, 2005, and June 30, 2007, shows that U.S. citizens received more than 95 percent of bachelor's degrees and more than 81 percent of master's degrees. An analysis of the 2009 Survey of Earned Doctorates,[7] which captures information on individuals receiving research doctorate degrees from an accredited U.S. institution during the 2009 academic year, shows that 94 percent of degrees conferred to U.S. citizens and permanent residents were conferred to U.S. citizens. In addition, some permanent residents will become eligible for naturalization, and thus for NGA positions, if they have been a permanent resident for at least 5 years.[8]

The education data show that more than 176,000 bachelor's degrees and more that 32,000 master's degrees in fields of study that are highly relevant to the core and emerging areas were conferred to U.S. citizens and permanent residents in 2009 (Table C.7, Appendix C), with annual growth rates of 3.5 percent for bachelor's degrees and 4.2 percent for master's degrees over the 2000–2009 period. As illustrated in Figure 4.5, the percentage of these degrees conferred to U.S. citizens and permanent residents was stable over the 2000–2009 period. Approximately 98 percent of bachelor's degrees and 82 percent of master's degrees were conferred to U.S. citizens and permanent residents during this period (Table C.8, Appendix C). In contrast, the annual growth rate of doctorate degrees conferred to U.S. citizens and permanent residents over the 2000–2009 period was a modest 0.8 percent, and the fraction of doctorate degrees conferred to U.S. citizens and permanent residents was lower in the second half of the decade. About 72 percent of doctorate degrees went to U.S. citizens and permanent residents in 2000 and 65 percent in 2009. Compared to the average for all core and emerging areas (Figure 4.5), there are notably fewer U.S. citizens and permanent residents in master's and doctorate programs that are highly relevant to crowdsourcing and in doctorate programs

that are highly relevant to geodesy and geophysics (Table C.9, Appendix C). Given that NGA hires individuals mainly at the bachelor's and master's levels, the number of U.S. citizens with doctorates may not be important to NGA.

To estimate the citizenship status of individuals already employed in relevant occupations, the committee used data collected in the American Community Survey, an ongoing survey conducted by the U.S. Census Bureau. Specifically, the 2005–2009 Public Use Microdata Sample Files were used to estimate the fraction of the workforce in NGA-relevant occupations that were U.S. citizens. The data show that at least 75 percent of employees in NGA-relevant occupations were U.S. citizens; in most of these occupations, more than 90 percent of employees were U.S. citizens (see Table D.5, Appendix D). In 2009, the occupations with the highest fraction of U.S. citizens were librarians (98 percent); surveying and mapping technicians (97 percent); statistical assistants (97 percent); aerospace engineers (97 percent); urban and regional planners (97 percent); and surveyors, cartographers, and photogrammetrists (96 percent). Occupations with the lowest fraction of U.S. citizens were physical scientists, all other (76 percent); computer software engineers (80 percent); and astronomers and physicists (86 percent).

Overall, the education and census data suggest that NGA's U.S. citizenship requirement does not dramatically reduce the "upper-bound" pool of qualified new graduates or experienced employees. When citizenship is factored in, the number of new graduates at all levels decreases by 7 percent to 214,870, and the number of experienced workers decreases by 12 percent to 2,417,964.

ANTICIPATED AVAILABILITY OF EXPERTS

Estimating the future availability of expertise in any field is inherently difficult and the estimates are subject to large uncertainties that grow with the projection horizon. The Bureau of Labor Statistics makes 10-year employment projections, but the projections reflect the anticipated demand for workers, not the supply of workers, in a given occupation.[9] Accurate forecasts of the number of graduates can be made for

---

[6] See <http://www.nsf.gov/statistics/srvyrecentgrads/>.

[7] See <http://www.nsf.gov/statistics/srvydoctorates/>.

[8] U.S. Citizenship and Immigration Services, A Guide to Naturalization, Chapter 4: Who is eligible for naturalization?, p. 18, <http://www.uscis.gov/portal/site/uscis>.

---

[9] See Bureau of Labor Statistics Employment Projections Program, <http://www.bls.gov/emp/>.

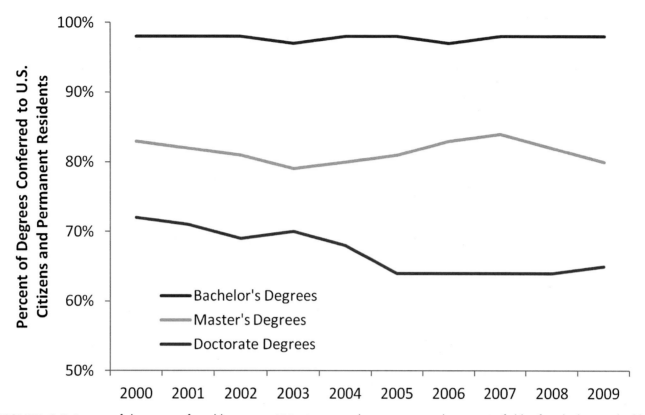

FIGURE 4.5 Percent of degrees conferred by year to U.S. citizens and permanent residents across fields of study that are highly relevant to the core and emerging areas. SOURCE: Data from the U.S. Department of Education, National Center for Education Statistics' Integrated Postsecondary Education Data System Completions Survey. Accessed via WebCASPAR.

a few years into the future because the future graduates are already attending classes and attrition rates are reasonably well known (NRC, 2000). As the projection horizon lengthens, however, the projections become less reliable as a result of (a) misspecification of models, (b) data that are flawed or aggregated at an inappropriate level, and/or (c) unanticipated events. For example, the profound effect of the September 11, 2001, terrorist attacks on scientific labor markets would have been difficult to incorporate into any forecasting model. Future labor markets depend on the paths of multiple variables, including some that are well understood (e.g., the age structure of the population) and others that are unpredictable (e.g., the career preferences of future college students, future technology changes that will affect the demand for talent, immigration policies). Given the uncertainties in labor forecast models as well as the resource constraints of an NRC study, the committee chose to estimate the future availability of geospatial intelligence expertise by simply extrapolating recent

trends in the number of graduates in relevant fields of study, as described below.

Supply Outlook

The future supply of geospatial intelligence expertise depends primarily on the number of people graduating with degrees in relevant fields of study. The committee projected the future supply of new graduates by extrapolating trends in the number of degrees conferred to U.S. citizens and permanent residents over the past decade to 2030. In its projections, the committee assumed that the annual growth rates in degrees conferred observed over the 2000–2009 period (i.e., 3.5 percent per year for bachelor's degrees, 4.2 percent per year for master's degrees, and 0.8 percent per year for doctorate degrees) will continue. Based on this assumption, the number of bachelor's degrees conferred in geospatial intelligence-related programs would be expected to climb to 367,000 by 2030, the number of master's degrees conferred would be expected to exceed 75,000, and the number of doctorate degrees conferred would be expected to rise to 6,000 (left column of Table 4.2).

TABLE 4.2 Projected Number of Degrees in Geospatial Intelligence-Related Fields of Study Conferred to U.S. Citizens and Permanent Residents in 2030

Degree Level	Scenario		
	Reference	High Growth	Low Growth
Bachelor's degrees	367,000	525,000	256,000
Master's degrees	77,000	117,000	50,000
Doctorate degrees	6,000	7,000	6,000
TOTAL	451,000	648,000	312,000

NOTE: Results are rounded to the nearest thousand. The projections were made using the standard future value formula (Brealey and Myers, 1996): beginning value $\times$ $(1+g)^N$, where beginning value is the number of degrees conferred in 2009, g is the observed annual growth rate of degrees conferred over the 2000–2009 period, and N is the number of years that will elapse between the beginning value year and the projection year (21 years).
SOURCE: Projections were made based on data from the U.S. Department of Education, National Center for Education Statistics' Integrated Postsecondary Education Data System Completions Survey. Accessed via WebCASPAR.

To illustrate the uncertainty associated with these forecasts, the committee extrapolated the number of new graduates under a high-growth scenario (50 percent higher than the growth rate observed for 2000–2009) and a low-growth scenario (50 percent lower than the observed growth rate). The results of these three scenarios are presented in Table 4.2 and illustrated in Figure 4.6. The number of degrees projected to be conferred in 2030 to U.S. citizens and permanent residents in geospatial intelligence-related fields of study ranges from 312,000 to 648,000. However, the actual values could be higher or lower, especially if any large shocks in the labor market occur over the next 20 years (e.g., NRC, 2000).

The projections presented above are based on trends in the total number of graduates with degrees that are highly relevant to the core and emerging areas. Only a small fraction of these graduates will have the combination of knowledge and skills suited for a science or analysis position at NGA. Thus, the projections are an "upper bound" on the future availability of expertise in geospatial intelligence.

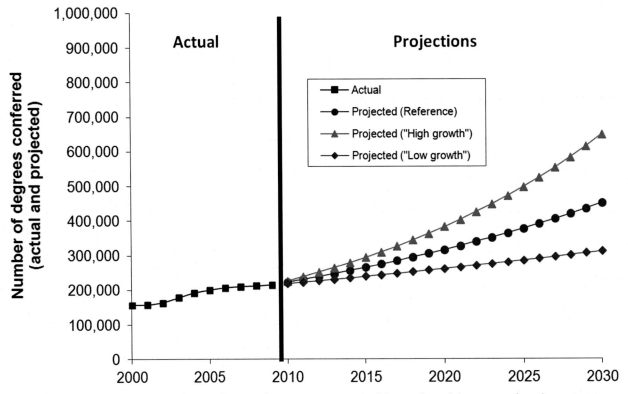

FIGURE 4.6 Observations (2000–2009) and extrapolations (2010–2030) of the number of degrees conferred to U.S. citizens and permanent residents by year across highly relevant geospatial intelligence-related fields of study. SOURCE: Data for 2000–2009 are from the U.S. Department of Education, National Center for Education Statistics' Integrated Postsecondary Education Data System Completions Survey. Accessed via WebCASPAR.

The committee's best estimate of the current number of experts (Table 4.1), which is considerably lower than the total number of graduates, does not lend itself to extrapolation because the numbers are difficult to aggregate and previous growth rates cannot be estimated with any certainty. However, based on the evolution of the core and emerging areas (Chapters 2 and 3), it is likely that the number of graduates in cartography and photogrammetry will decline and that the number of graduates in other areas, especially the emerging areas, will grow over the next 20 years. Thus, although the exact number cannot be projected with high confidence, it is likely that the supply of graduates in all geospatial intelligence-related fields of study except photogrammetry and cartography will be robust for the next 20 years.

SUMMARY AND CONCLUSIONS

The first task of the committee was to estimate the number of experts in the core and emerging areas now and over the next 20 years. The primary sources of expertise are new graduates in relevant fields of study, which are tracked by the Department of Education, and employees working in occupations that require relevant knowledge and skills, which are tracked by the Bureau of Labor Statistics. Unfortunately these data are not ideal for addressing the task because the core and emerging areas are embedded within or span multiple fields of study or occupations; 109 fields of study and 36 occupations potentially provide some knowledge and skills relevant to a core or emerging area.

An "upper bound" on the number of experts was determined by summing the number of new graduates from all relevant fields of study and the number of workers from all relevant occupations. This calculation showed that there were more than 229,000 degrees conferred in the 109 instructional programs in 2009 and more than 2.7 million jobs in the 36 occupations in 2010, the latest years for which data were available when the report was being written. NGA's requirement for U.S. citizenship decreases the size of the labor pool by 7 percent for new graduates, with the largest reductions at the doctorate level, and by 12 percent for experienced workers, with the largest reductions in physical science and computer occupations. Accounting for U.S. citizenship reduces the estimate to more

than 200,000 new graduates in relevant instructional program in 2009 and more than 2.4 million jobs in relevant occupations. These graduates and experienced workers likely have some knowledge and skills in a core or emerging area and could potentially be trained for an NGA position.

The actual number of graduates with expertise in a core or emerging area is likely considerably lower than the "upper-bound" estimates, especially for the emerging areas, which are taught as comprehensive programs in only a handful of universities. Factoring in other information—including the number of universities offering programs in a core or emerging area, the size of the professional community, and the number of graduates from instructional programs that produce the bulk of necessary skills—allows a qualitative estimate to be made based on expert judgment. The committee's best estimate is that the current number of new graduates in geospatial intelligence areas is likely on the order of tens for photogrammetry; tens to hundreds for GEOINT fusion, crowdsourcing, human geography, and visual analytics; hundreds for geodesy, geophysics, and cartographic science; hundreds to thousands for remote sensing and forecasting; and thousands for GIS and geospatial analysis.

Insufficient information was available to refine the number of experienced workers. A "lower bound" was estimated by summing the number of jobs in the four most closely related occupations: cartographers and photogrammetrists; surveying and mapping technicians; geographers; and geoscientists, except hydrologists and geographers. For these four occupations, there were nearly 100,000 jobs in 2010.

Estimates of the future availability of experts are subject to large uncertainties, so the committee simply extrapolated past trends of the number of new graduates with relevant degrees. Extrapolation of 10-year trends under high-growth and low-growth scenarios suggests that 312,000–648,000 degrees in relevant fields of study will be conferred to U.S. citizens and permanent residents in 2030. These figures place an "upper bound" on the future number of graduates in geospatial intelligence-related fields of study. The committee's best estimate of the current number of graduates with skills and knowledge in core and emerging areas is qualitative and could not be projected with confidence. However, it is substantially lower than the total number of graduates in relevant fields of study and thus would yield substantially lower projections. Based on how the core and emerging areas have evolved over the past few decades, it is likely that the number of graduates will rise in all areas except photogrammetry and cartography over the next 20 years.

Based on the education and labor analysis, the answer to Task 1 is that the current number of U.S. citizens and permanent residents with education is likely on the order of tens for photogrammetry; tens to hundreds for GEOINT fusion, crowdsourcing, human geography, and visual analytics; hundreds for geodesy, geophysics, and cartographic science; hundreds to thousands for remote sensing and forecasting; and thousands for GIS and geospatial analysis. In addition, U.S. citizens currently hold more than 100,000 jobs in occupations closely related to the core areas. If substantial on-the-job training is an option for NGA, the current labor pool increases to 200,000 new graduates and 2.4 million experienced workers. If 10-year growth trends in the "upper-bound" estimate continue, the number of new graduates could reach 312,000–649,000 by 2030.

5

Current and Anticipated Gaps in Expertise

Chapters 2 and 3 described the knowledge and skills required for a position in a core and emerging area, and Chapter 4 provided estimates of the number of experts (new graduates and experienced workers) in these areas. This chapter compares these results with information on the National Geospatial-Intelligence Agency's (NGA's) needs to identify gaps in the current or future availability of geospatial intelligence expertise (the committee's Task 2). The committee examined gaps in domain knowledge and skills and where to find them. NGA's current needs were estimated from information provided by the agency (see Box 1.3) or available on its website. In particular, the job listings[1] and occupation descriptions for scientists and analysts (Appendix B) provide a measure of the knowledge and skills the agency is currently seeking, and the schools where NGA recruits potential employees indicate where the agency is looking for this knowledge and skills. The curriculum of the NGA College was assumed to reflect not only what topics are currently important to the agency, but also what knowledge and skills are hard to find in new employees.

Estimating NGA's needs over the next 20 years is more difficult, in part because trends in hiring may have changed. Moreover, ongoing scientific and technological advances and evolving needs for geospatial intelligence continually change the skill sets needed. In addition, the bimodal age distribution of NGA's scientists and analysts (Box 5.1) means that junior staff likely have different skills and analysis workflows than those

nearing retirement. As these staff move into leadership positions, the agency culture will change, possibly attracting new recruits or accelerating the departure of some staff (see Wilkins and Ouchi, 1983; Carley, 2000; and Cameron and Quinn, 2006, for a discussion of changing organizational cultures). The cultural shift will also change what technologies are used and what skills are sought. Finally, the beginning of the age of big data (Manyika et al., 2011) and ubiquitous geospatial information are driving rapid growth in the geospatial industry as well as creating more competition for graduates with geospatial knowledge and skills (e.g., Gewin, 2004; DiBiase et al., 2006; Solem et al., 2008). The impacts of these changes are difficult to forecast, so the committee estimated NGA's future needs based on the age distribution of NGA's current geospatial intelligence workforce and the assumption that future hiring would focus on the core and emerging areas.

DOMAIN KNOWLEDGE

The Chapter 4 education and labor analysis yielded estimates of the number of new graduates with education in the core and emerging areas, as well as estimates of the number of experienced workers in closely related occupations. NGA generally hires several hundred people from these two sources each year. Below we compare the education and labor estimates with NGA's needs for domain knowledge in the core and emerging areas over the next few decades.

---

[1] See <https://www1.nga.mil/CAREERS/CAREEROPP/ Pages/default.aspx>.

BOX 5.1
Age Distribution of NGA
Scientists and Analysts

The success of recruitment during the years following the September 2001 terrorist attacks in New York and Washington, D.C., led to a bimodal age distribution of NGA scientists and analysts. Compared to other federal agencies, NGA has a relatively young workforce, with only a small fraction of scientists and analysts over 60 years old. If current staff retire at age 65, the first major round of retirements will begin by the end of the decade.

SOURCE: NGA.

Core Areas

More than half of geospatial intelligence analyst positions at NGA specify degrees or coursework in Geographic Information Systems (GIS), geospatial analysis, geography, or geographic information science (Table B.1, Appendix B). Approximately 189 universities offer relevant degrees, and hundreds of community colleges offer relevant courses (Table A.5, Appendix A). In 2009, 5,404 U.S. citizens and permanent residents received a degree in geography, the instructional program that produces the bulk of expertise in GIS and geospatial analysis (Table C.10, Appendix C). The number of geography graduates far exceeds the number of geography jobs nationwide (1,300 jobs in 2010; see Table D.2, Appendix D) and the field is growing, suggesting that the supply of geographers will be sufficient for NGA's needs over the next 20 years. On the other hand, GIS applications analysts are in high demand by the private sector, with qualified candidates difficult to find (Mondello et al., 2004, 2008; Solem et al., 2008). Given that the NGA College offers reasonably comprehensive coursework in GIS operations (Box 5.2), it is possible that competition from private companies is already making it difficult to find qualified experts in GIS applications and techniques.

Expertise in remote sensing is also important to NGA: remote sensing appears in the education requirements for nearly half of NGA scientist and analyst occupations (Table B.1, Appendix B), and a few thousand NGA scientists and analysts work on imagery analysis. The supply of remote sensing gradu-

ates is likely on the order of hundreds to thousands (Table 4.1). The supply of experienced workers in the most closely related occupation (physical scientists, all others) is 24,690 (Table D.2, Appendix D). Although the supply exceeds the number of NGA positions, the NGA College places heavy emphasis on remote sensing (Box 5.2), suggesting that extensive on-the-job training is already necessary for remote sensing and imagery analysis positions.

Compared to GIS and remote sensing, a relatively small number of NGA positions require specialized knowledge in cartography, geodesy and geophysics, or photogrammetry. A bachelor's degree in cartography or at least 30 semester hours of cartography coursework is required for NGA analyst positions in cartography and photogrammetry (Table B.1, Appendix B). Only 155 U.S. citizens or permanent residents obtained a degree in cartography in 2009 (Tables C.6 and C.10, Appendix C), but there is a large supply of cartography and photogrammetry professionals (11,670), working mainly in the private sector (Table D.2, Appendix D). The NGA College offers minimal training in cartography (Box 5.2), suggesting that NGA is currently able to find enough qualified candidates. However, the agency is likely to face a shortage (i.e., numbers are too small to give NGA choices or means of meeting sudden demand) in the near future. Employer surveys have identified cartographers as among the most difficult positions to fill (Mondello et al., 2004, 2008; Solem et al., 2008). Moreover, cartography appears to be losing its identity as an academic discipline. Fewer colleges and universities offer degrees or certificates in cartography, and more students are choosing instead to pursue a specialization in geographic information science, remote sensing, or spatial analysis (see Chapter 2).

The situation is worse for photogrammetry, which has nearly disappeared as a field of study in academia. Only 15 universities offer photogrammetry classes (Table A.2, Appendix A), and only 26 U.S. citizens or permanent residents obtained a degree in a closely related field (surveying engineering) in 2009 (Table C.10, Appendix C). A degree in photogrammetry is not required for any NGA position, but coursework in photogrammetry is identified as useful for several occupations, including those related to photogrammetry, cartography, geodesy, and data col-

---

BOX 5.2
NGA College

The NGA College is an accredited institution housed within NGA that offers approximately 170 courses in geospatial intelligence, leadership, and professional development to government civilians, members of the military, and contractors to NGA and other U.S. defense and intelligence agencies.[a] The specific training required for new employees depends on the requirements of the position, along with the skills, education, and experience of the individual.[b] Classes are taught by government employees and contractors[c] and typically last between 1 and 5 days. The longest class, basic geographic intelligence, runs about 7 months. About 15,000 students receive training in the college each year.

Nearly 40 percent of the classes offered at the college are related to remote sensing and offer a reasonably comprehensive suite of classes in data collection strategies, image processing, and major remote sensing systems, including infrared, multispectral/hyperspectral, radar/polarimetry, and motion imagery. The treatment of GIS operations using commercial products is also reasonably complete, but there is little coursework in geospatial analysis, such as spatial data analysis, spatial statistical analysis, or spatial optimization. None of the courses focus on geospatial data visualization and information design, even though NGA cartographers and other analysts work with graphics, imagery, movies, and maps.

Classes relevant to other core areas are sparse and introductory in nature. For example, no geophysics classes are offered. A few courses teach basic geodesy concepts; none deal with more advanced concepts, such as platform navigation, charting, Global Navigation Satellite Systems such as the Global Positioning System, or mathematics or statistics. Similarly, the only class offered in photogrammetry is taught at the introductory level, although some photogrammetric concepts, theory, procedures, exploitation techniques, and product quality issues are taught in the remote sensing courses.

Not surprisingly, the emerging areas are poorly covered in the current NGA College curriculum. For example, a few courses touch on methods to visually overlay disparate data, but none cover broader GEOINT fusion concepts such as ontology, the semantic web, schema integration, map conflation, or statistical methods of combining different types of evidence. Similarly, a few courses offer basic information useful to visual analytics (e.g., Google Earth and related applications) and to intelligence forecasting or scenario forecasting. Although two courses mention network analysis, the subtopics of strong relevance to NGA (dynamic network analysis and geospatial network analysis) are not covered. No courses discuss the use and limitations of crowdsourcing for creating maps and gathering data, although some of the relevant technologies (e.g., Google Earth, text mining) are covered.

[a] See <https://www1.nga.mil/MediaRoom/Publications/Documents/Factsheets/NCE_College.pdf>.
[b] See <https://www1.nga.mil/NGAJobs/Pages/Occupations.aspx>.
[c] Presentation to the committee by Mark Pahls, Chief of Learning Integration, NGA College, on May 23, 2011.

---

lection, as well as to principal and project scientists (Table B.1, Appendix B). The NGA College offers only one introductory course in photogrammetry (Box 5.2), suggesting that qualified candidates are currently available. Much of the stock of trained photogrammetry professionals resides in private companies, including contractors to NGA. There were more than 7,000 jobs in cartography and photogrammetry in the private sector in 2010 (Table D.2, Appendix D). Although this source of experts may be sufficient for NGA's needs in the short run, the lack of rigorous university training in photogrammetry will eventually yield a shortage of photogrammetrists qualified for a position at NGA.

Geodesy-related positions at NGA require a bachelor's degree in geodesy, mathematics, physical science, or a related discipline (Table B.1, Appendix B). NGA has no specific positions in geophysics (or courses at the NGA College; Box 5.2), although coursework or experience in geophysics is identified as useful for cartography, geodesy, photogrammetry,

and principal and project scientist positions. In 2009, 138 U.S. citizens and permanent residents received a degree in geophysics and seismology, and 26 received a degree in surveying engineering (Table C.10, Appendix C), the instructional programs that produce the most geophysicists and geodesists. Much larger numbers of experts were employed in 2010, including more than 30,000 geoscientists and more than 50,000 surveying and mapping technicians (Table D.2, Appendix D), the most closely related occupations. This supply is large relative to NGA's current needs. However, the supply of graduates is small (on the order of hundreds) and only about one-third of these have advanced degrees and specialized training in geodesy. The small number of geodesy graduates, coupled with federal agency concerns about a growing deficit of highly skilled geodesists (NRC, 2010c), suggests that NGA may soon have to hire and train professionals from other disciplines. Indeed, the few geodesy-related courses at the NGA College appear to be geared toward analysts trained in other disciplines.

Emerging Areas

NGA currently has no science or analyst positions in the emerging areas, although some of the knowledge relevant to human geography is needed for NGA analyst positions in political geography, regional geography, regional source, and scientific linguistics. Consequently, any gaps in the supply of expertise in the emerging areas relative to NGA's needs will occur in the future. It is likely that NGA's need for expertise in the emerging areas will grow over time. The increasing availability of geospatial data and technology are allowing NGA to tackle increasingly complex intelligence problems, which commonly require interdisciplinary approaches (Box 5.3), such as those embodied in the emerging areas.

By their nature, training in the emerging areas is provided through individual courses often scattered among different university departments. Each program has a unique set of collaborating departments and approach for dealing with the topics, which creates difficulties for finding expertise. For example, different departments tend to explore different aspects of fusion, leading to multiple (and sometimes inconsistent) vocabularies and conceptualizations. Much of the technology development for human geography takes place in computer science, electrical engineering, and physics departments without reference to the large body of theoretical and empirical work in geographic and social science departments, leading to an increasing divergence between theory and methods. The lack of standard curricula, established journals, and even a common language means that graduates from different programs will have different knowledge and skills.

The other major gap associated with the emerging areas is the number of graduates. Fewer than a dozen universities offer specialized training in any emerging area except forecasting, and only a few universities offer a comprehensive degree program (Chapter 3). Anecdotal evidence suggests that many of these graduates are finding jobs quickly,[2] so competition, combined with a small supply (tens to hundreds in most emerging areas; see Table 4.1), could lead to shortages in the future availability of expertise in the emerging areas.

---

[2] The emerging areas can be considered data science jobs— those requiring expertise in multiple technical disciplines, such as computer science, analytics, math, modeling, and statistics. Such jobs are expected to see a shortage of 190,000 data scientists by 2018 (Bertolucci, 2012).

BOX 5.3
Interdisciplinary Approaches

A common critique of disciplinary science is that it leads practitioners to look inward and to create numerous subspecialties in what scholars have called the fragmentation of disciplinarity (Strober, 2006) or stovepiping. Countering this tendency is the more recent recognition that scientific breakthroughs often happen at the edges and intersections of disciplines and specialties (Kates, 1987). These intersections occur at a range of scales. Multidisciplinary approaches involve people with different skill and knowledge sets working together, such as a geodesist working with a cartographer as part of a geospatial intelligence team, and they require an infrastructure for information sharing, such as a control room or social network. Interdisciplinary approaches require people to train across multiple fields (e.g., astrobiology). People with interdisciplinary skills may act as catalysts to problem solving, particularly when no approach seems suitable within an existing discipline (e.g., Omenn, 2006). Finally, transdisciplinary research problems are too large and complex to solve by any one discipline (Jantsch, 1972). Examples of transdisciplinary projects include climate change research, mapping the human genome, and testing the laws of physics using the Large Hadron Collider.

Few universities have succeeded in training interdisciplinary students because college and departmental structures often discourage the approach, and only a handful have mastered multidisciplinary approaches. Once created, interdisciplinary programs are hard to maintain because peer-review processes are commonly organized along traditional discipline lines. Most interdisciplinary training takes place at the graduate level. However, undergraduate students can achieve these goals by choosing double majors; multiple minors; and interdisciplinary, self-guided, and mixed-mode majors. For example, many students study abroad, create internships, do voluntary work, and seek out accreditation and certificate programs. Such combinations may eventually outnumber more traditional majors.

SKILLS

The distinction between knowledge and skills is not always clear, especially for the geospatial field, which can be viewed as a discipline, a collection of tools, or a profession (DiBiase et al., 2006, 2010). In 2010, the Department of Labor's Employment and Training Administration issued a geospatial technology competency model to define the scope of disciplines and the training and credentials required to work in the geospatial technology industry. The model lays out tiers of competencies, or capabilities for using sets of knowledge, skills, and abilities to successfully perform specific tasks (Figure 5.1). Tiers 1–3 describe general workplace behaviors and knowledge needed in most industries, including personal attributes learned at home (e.g., interpersonal skills, integrity), knowledge and skills learned in academic settings (e.g., geography, communication, basic computer skills), and skills honed in the workplace (e.g., teamwork, creative thinking). Tier 4 describes subjects (e.g., remote sensing, GIS, programming) and background knowledge (e.g., analytical

methods, geospatial data) needed by many geospatial professionals in their careers. Tier 5 specifies clusters of subject and background knowledge needed for each of three industry sectors: positioning and geospatial data acquisition; analysis and modeling; and software and application development. Above these tiers are competencies required for specific occupations (e.g., cartographers and photogrammetrists) and managers.

NGA occupation descriptions specify a set of core competencies for all science and analyst positions as

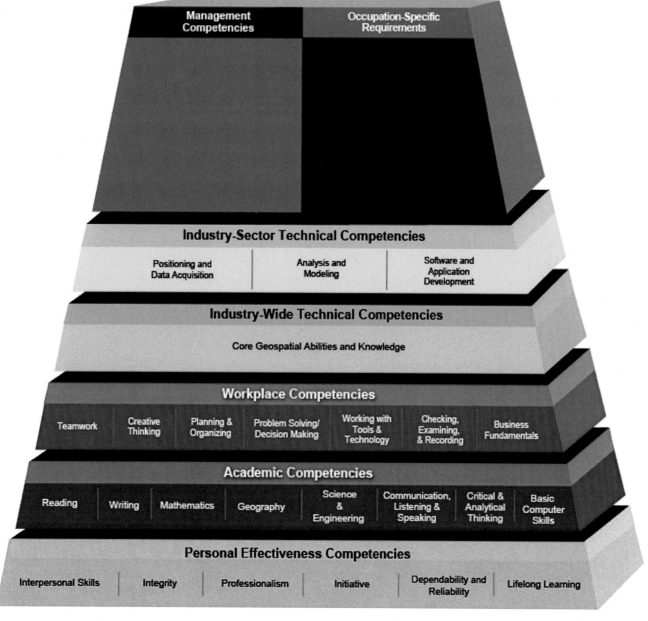

FIGURE 5.1 Geospatial technology competency model. SOURCE: Department of Labor, <http://www.careeronestop.org/competencymodel/pyramid.aspx?GEO=Y>.

well as the skills required for each type of position. The core competencies and skills span all levels of the geospatial technology competency model, although the core competencies stress interpersonal skills, communication, and creative thinking and adaptability, whereas the position-related skills stress working with customers and gathering, analyzing, and disseminating information. The most common skills among NGA science and analysis positions are illustrated in Figure 5.2.

The NGA College offers several courses in interpersonal skills, effective communication, and critical thinking, suggesting that these core competencies are in short supply. These skills are taught in some university programs, and new ways of teaching may also help fill the gap. For example, techniques such as role playing, gaming, and self-assessment favor understanding and conceptual methods, rather than content and memorization.

In the foreseeable future, new questions, as well as the data sets and tools needed to answer them, will continually arise. Dealing with these evolving questions and approaches requires a flexible workforce that is capable of thinking in breadth, rather than depth, through interdisciplinary training and teamwork. Historically, NGA employees acquired the necessary breadth of skills through an undergraduate education in a relevant discipline, internships or service, and/or training through the NGA College. However, the in-

creasing demands of teamwork and of multidisciplinary and interdisciplinary analytical tasks are placing increasing importance on a broader set of skills.

If there were such a thing as an ideal geospatial intelligence analyst, he or she would be well versed and expert at spatial thinking; have considerable interdisciplinary training; be well traveled and knowledgeable of world cultures (and able to use tools such as Google Earth for rapid virtual travel); have some core background in statistics, cartography (coordinates, projections, scale), and computer science (programming principles, operating systems); have a high degree of science literacy; read and write multiple languages; and have a commitment to professional ethics. None of these skills are classified as core competencies of NGA scientists and analysts, and skills in statistics, ethics, cultural analysis, and scientific methods are required only for certain NGA positions. Consequently, it is likely that NGA scientists and analysts are missing skills that will be important for future work in the core and emerging areas.

University departments commonly teach some of these skills. Spatial literacy and spatial reasoning are finding their way into undergraduate and graduate geography curricula nationwide (NRC, 2006). Spatial thinking is highly interdisciplinary, an extension of efforts to bring methods from GIS and spatial analysis into the social sciences and humanities. For example,

FIGURE 5.2 Word cloud illustrating the 25 most common skills identified in job descriptions for NGA scientists and analysts in the five core areas. The most common topics are portrayed by the largest lettering, and an arbitrary color scheme is used to distinguish the various phrases. SOURCE: Generated using <http://www.wordle.net>.

the Center for Spatially Integrated Social Science[3] was a 5-year National Science Foundation project designed to expand the knowledge and use of GIS and spatial methods in the social sciences, including demography, sociology, landscape architecture, and other disciplines.

Computer programming skills are needed for many of the core and emerging areas. For example, dealing with big data in geospatial intelligence (GEOINT) fusion, forecasting, visual analytics, and human geography requires skills in database management and construction for large data sets, natural language processing and text mining for large text data streams, social media mining, and streaming image or video processing. These skills are generally learned in computer science, information systems, or information technology programs. Even when data volumes are modest, computer programming skills are needed for writing scripts to encode image analysis and processing steps, implementing algorithms, understanding methods such as tracking and optimization, and communicating effectively with programming staff.

Other skills required for most of the core and emerging areas include statistics, network theory, and advanced mathematics. However, many geography departments, where the cartography and geographic information science specializations are commonly housed, no longer require calculus, statistics, or basic programming, and they have never required network theory. Students in geography do not naturally drift toward coursework in these areas, and it is difficult to teach someone to map residuals, for example, when he or she does not understand means and variance, root-mean-square error, or even the difference between a standard deviation and an interquartile range. Engineering and computer science students have some of this training (particularly in computer programming and advanced mathematics), but they generally have few spatial skills.

Similarly, advanced quantitative skills are required for forecasting, which is based on analog (e.g., similar patterns), analytical (e.g., physical or mathematical), statistical (e.g., deterministic, stochastic), or computational (e.g., numerical models, data-model assimilation) methods. Geospatial forecasting needs to connect components and interactions from physical, social, and

cultural systems. However, the majority of GIS or social science students lack adequate mathematical capabilities for geospatial forecasting, although the number of social science students in programs that emphasize statistics, agent-based modeling, and social networks is growing. Students in the physical, environmental, or life sciences generally have better quantitative skills, but they lack the abilities to handle the diverse, uncertain, and culturally and geographically dependent nature of the human dimension.

Other quantitative methods useful to many of the core and emerging areas include visualization and graphics design, modeling and simulation (usually left for graduate school), and the analysis of geospatial data from social media. For example, the suite of software commonly used by students has broadened from standard statistical packages and GIS to include visual analytics, semantic web, content analysis, and others. Standard, often commercial packages have rapidly yielded to extendable "mashups" of open-source software, although few university programs take advantage of this rapid expansion in the type and nature of analytical tools.

Finally, students commonly lack capabilities in the qualitative methods (e.g., interviews, questionnaires, textual content analysis, ethnographic assessment) that are often needed in addition to the quantitative methods discussed above. Few programs teach these methods, despite their importance to many research fields.

Overall, changes in university programs are making some skills needed by NGA scientists and analysts harder to find (e.g., cartographers with math and programming skills) and others easier to find (e.g., geographers with spatial thinking skills). The emergence of interdisciplinary areas such as GEOINT fusion, visual analytics, and human geography is beginning to yield graduates with skills from several university departments (e.g., computer science and spatial skills). However, until these programs develop, individuals with the ideal combination of skills for NGA are likely to remain in short supply.

RECRUITING

NGA focuses recruiting on dozens of colleges and universities that are near major NGA facilities (i.e., Springfield, Virginia; Saint Louis, Missouri) or that

---

[3] See <http://www.csiss.org/>.

have a large population of underrepresented groups (e.g., historically black colleges and universities). Few of these institutions have significant programs in core or emerging areas, although they likely meet other agency goals, such as increasing diversity. About one-third of the schools and universities where NGA recruits are large state universities, and several of these (e.g., George Mason University, Ohio State University, Pennsylvania State University, University of California, Santa Barbara) offer education and training in several core or emerging areas. Extending recruiting to some of the example universities listed in this report (e.g., Tables A.1–A.11, Appendix A) would help NGA find individuals with knowledge and skills in core and emerging areas.

SUMMARY AND CONCLUSIONS

The second task of the committee was to identify gaps in the current or future availability of expertise relative to NGA's needs. The Chapter 4 analysis showed that the number of new graduates with education in core and emerging areas and the number of experienced workers in closely related occupations far exceeds NGA's needs for expertise in all core and emerging areas (generally several hundred people a year). However, when other considerations are factored in—including competition from other organizations and the extensive training provided by NGA in some areas—a more nuanced picture emerges. Expertise in geophysics and geospatial analysis is likely sufficient for NGA's current and future needs. NGA hires only a small fraction of the available experts and offers little or no training in these areas to employees through the NGA College. The supply of experts in cartography, photogrammetry, and geodesy appears adequate for now. The number of professionals working in these areas is substantially higher than the number of NGA job openings, and only minimal training is offered at the NGA College. However, some shortages are likely in the future because photogrammetry, geodesy, and cartography programs produce a small number of graduates, and the number of academic programs in photogrammetry and cartography is shrinking. Moreover, employer surveys suggest that skilled cartographers and geodesists are hard to find. Shortages may already be appearing in GIS and remote sensing, given the extensive training

in these fields provided by the NGA College. Although the supply in both fields exceeds NGA's needs, competition for GIS applications analysts is strong. By definition, NGA has no current positions for experts in emerging areas, but as the agency tackles increasingly complex geospatial intelligence problems, demand for the types of interdisciplinary approaches embodied by the emerging areas is likely to grow.

In addition to domain knowledge and interdisciplinary skills, NGA scientists and analysts need a variety of personal, academic, and workplace skills. The NGA College offers several courses in interpersonal skills, written and oral communication, and critical thinking, suggesting that these skills are currently in short supply. In NGA's future workforce, which is likely to be more interdisciplinary and focused on emerging areas, the ideal skills will include spatial thinking, scientific and computer literacy, mathematics and statistics, languages and world travel, and professional ethics. These skills are not always taught in university programs. Although spatial thinking is increasingly being taught in undergraduate programs, math and computer skills remain a gap in many natural and social science programs, and spatial perspectives remain a gap in most computer science and engineering programs.

Individuals with the knowledge and skills needed for a geospatial intelligence position at NGA are available, but NGA may not be looking for them in all the right places. Only about one-third of the universities and colleges where NGA currently focuses recruiting have strong programs in core or emerging areas. The academic institutions discussed in this report may provide a useful start for finding programs in core and emerging areas.

In summary, the analysis for Task 2 revealed both current and future gaps in knowledge and skills relative to NGA's needs. Although the supply of experts is larger than NGA demand in all core and emerging areas, competition may be making GIS and remote sensing experts hard to find. Long before 2030, competition and a small number of graduates will likely result in shortages in cartography, photogrammetry, geodesy, and all emerging areas. In NGA's future workforce, which is likely to be more interdisciplinary and focused on emerging areas, the ideal skill set will include spatial thinking, scientific and computer literacy, mathematics and statistics, languages and world culture, and professional ethics. Although NGA is currently finding employees with skills in statistics, ethics, cultural analysis, and scientific methods, graduates with the ideal skill set will remain scarce until interdisciplinary and emerging areas develop. NGA could improve its chances of finding the necessary knowledge and skills by extending recruiting to the example university programs identified in this report.

6

Current Training Programs

The gaps in geospatial intelligence-related knowledge and skills identified in Chapter 5 can be filled through education and training. Training in the disciplines, methods, and technologies underlying geospatial intelligence is offered by a variety of organizations. University undergraduate programs provide the basic knowledge and skills needed by National Geospatial-Intelligence Agency (NGA) scientists and analysts. More specialized training is available through university graduate programs, professional development programs sponsored by government agencies and private companies, and workshops and short courses offered by professional and scientific societies. This chapter describes education and training programs relevant to geospatial intelligence offered by these diverse organizations (Task 3). Few of these programs were designed specifically for NGA's employment needs and, thus, do not offer all of the knowledge and skills needed by the agency (e.g., mathematics, statistics).

The knowledge, skills, and techniques used to produce geospatial intelligence are diverse, so the list of relevant education and training programs is long. Consequently, the committee chose a set of representative examples in universities, government agencies, professional societies, and industry that meet a range of geospatial intelligence needs. Examples were chosen based on two or more of the following criteria:

- a long record of accomplishment in producing graduates with relevant knowledge and skills;
- a critical mass of high-caliber instructors;

- a significant number of students receive training; and
- an opportunity to solve problems in a real world context.

UNIVERSITIES

Universities provide students with a strong foundation in state-of-the-art geospatial science and technology as well as a means to augment skills with specialized training in a particular subject. Some universities also provide other types of skills and experience useful for producing geospatial intelligence, such as the ability to think and work across discipline boundaries, to combine scientific knowledge with practical workforce skills, or to apply scientific knowledge to solve real-world problems. Examples are described below.

Undergraduate Degree

Undergraduate degree programs are the primary supplier of geospatial skills, concepts, and knowledge for most geospatial analysts. The Department of Geography at the University of Colorado, Boulder, offers a typical undergraduate curriculum that teaches geospatial knowledge and skills. The department offers three primary emphases in geographic information science: Geographic Information Systems (GIS), remote sensing, and cartographic visualization. The GIS track focuses on spatial data structures and algorithms and on the application of GIS for modeling physical and human systems. The remote sensing track emphasizes image processing for environmental modeling and monitoring. The cartographic visualization track emphasizes geographic information design and Internet and World Wide Web spatial data applications.

The University of Colorado undergraduate major in geography with a concentration in geographic information science requires 45 credits. The flow of geography courses is shown in Figure 6.1 and descriptions of the courses are given in Table A.11 in Appendix A. The term "required" indicates which courses satisfy prerequisites in the flow. For example, GEOG 3023 (Statistics for Earth Sciences) is prerequisite

to GEOG 4103 (Introduction to GIS) and GEOG 4203 (Advanced Quantitative Methods). Core skills covered in these classes include the understanding of maps, scale, geodesy and map projections, cartographic transformations, georectification, and coordinate systems. More advanced classes cover map analysis, image analysis and interpretation, the representation of geographical objects and fields, spatial modeling, and statistics. The classes require students to download, merge, and process imagery and map data, and to create new versions and interpretations to suit a particular goal.

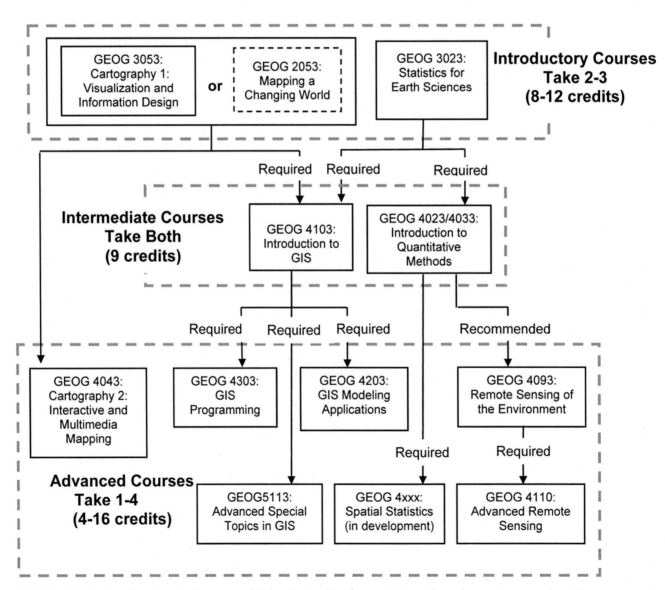

FIGURE 6.1 Course of study at the University of Colorado, Boulder, for an undergraduate degree in geography with a concentration in geographic information science. GEOG 2053 (dashed box) is optional.

Graduate Degree

Some universities offer graduate programs designed especially for federal workers and contractors. For example, the master's of science program at George Mason University is tailored to military or government contractors and emphasizes training to solve real-world problems. George Mason University is located near a cluster of federal employers in the Washington, D.C., region, which is attractive to workers seeking part-time graduate training while remaining employed. Moreover, many instructors come from the federal government, which facilitates communication between an academic training center and government employers.

The George Mason University master's in geographic and cartographic sciences focuses on preparing students for careers in geography, remote sensing, and GIS, as well as cartography, visualization, and modeling.[1] Courses concentrate on the techniques of compilation, display, and analysis of spatial data, and on associated applications (see Table A.12, Appendix A). In addition to a core curriculum consisting of coursework in GIS, remote sensing, and quantitative methods, students may take up to 24 credits of elective coursework. Electives focus on environmental applications, such as land use or hydrographic applications; on cultural and human systems, such as transportation, food security, and medical applications (e.g., GIS applications to model disease vectors); or on strategic applications, such as the geography of insurgency. In-depth training in cartography, spatial database management, and programming is also offered.

Professional Science Master's Program. Many employers are looking for individuals who possess scientific expertise along with practical workforce skills in communication, management, legal and regulatory affairs, and administration. Professional science master's programs are intended to produce graduates with this mix of scientific and practical skills. Internships are required of students, and an employer advisory board must be involved in program design and evaluation. The degree model was developed from an initiative of the Sloan Foundation in the 1990s, and professional science master's programs are now coordinated by the Council of Graduate Schools.[2]

Four universities offer geospatially oriented professional science master's programs under the Council of Graduate Schools. North Carolina State University offers a professional science master's degree in Geospatial Information Science and Technology, which is aimed at the development, management, and application of new technology to understand and manage spatial phenomena, such as economic development, disease, emergency planning and response, and environmental resources. Other relevant professional science master's programs include Applied Geospatial Sciences, Geospatial Technologies Emphasis (Northern Arizona University), Cartography and Geographical Information Systems (Binghamton University), and Geographical Information Systems and Remote Sensing (University of Pittsburgh).

Some master's programs have similarities to the professional science master's model, even though they lack the formal designation by the Council of Graduate Schools. For example, several universities offer professional master's degrees aimed at current practitioners as well as students seeking employment in the geospatial technology industry. Examples include the Pennsylvania State University, University of Minnesota, University of Southern California, Northwest Missouri State University, University of Denver, and the University of Colorado, Denver. Some master's programs in geography require internships, hire professionals to teach courses as adjunct instructors, and specify applied geography in the degree title or area of concentration.

Distance Learning

The traditional campus-based programs discussed above are not convenient for everyone, including individuals who work full time or who live far from a suitable university. Distance learning programs offer students a path to receiving formal training without requiring physical access to campuses as well as the flexibility to choose the time of day they devote to study. A growing number of universities provide on-

---

[1] See <http://ggs.gmu.edu/AcademicPrograms/MSGECA/MSGECAGuidelines.pdf>.

[2] See the program description provided by the National Professional Science Masters Association at <www.npsma.org> and the criteria required for a Professional Science Masters affiliation at <http://www.sciencemasters.com/>.

line education and training in geospatial areas. Pennsylvania State University was one of the first to offer such programs and its online GIS curriculum has had time to mature. The Pennsylvania State World Campus offers a number of professional degrees, without the requirement of a thesis, including a master's in geographic information systems and another in geospatial intelligence.[3] Courses are taught by 25 instructors, about half of whom are in residence at Pennsylvania State University. All courses are 10 weeks long and require 8 to 12 hours of student effort per week. The degree requires 35 credits, including 6 credits at the 400 level and a minimum of 12 additional credits at the 500 level or above (see course descriptions in Table A.13, Appendix A). Graduate-level courses in technical writing or project management may be counted toward the degree as electives. Topics covered include geodatabase design, spatial analysis, project management, and geospatial data analysis.

An interesting advance in distance learning is Massive Open Online Courses (MOOCs), which provide free online courses to anyone interested. Some MOOCs have attracted more than 100,000 students (Lewin, 2012). MOOCs have the potential to reach students who might not otherwise learn about geospatial science or technology. However, they have yet to be proven effective (e.g., Fini, 2009), and the specialized topics of interest to NGA may not attract the large numbers of students targeted by MOOCs.

Certificates

Certificate programs are a popular training option for students or professionals who want to augment their skills without obtaining another degree. Certificates acknowledge special training in selected subject matter, and they are commonly awarded as part of a bachelor's, master's, or doctorate degree. Obtaining a certificate does not mean that a person is legally "registered" in a profession. Registration usually requires the applicant to pass a rigorous standardized test administered by state-legislated authorities.

A substantial number of institutions grant certificates in cartography, GIS, and remote sensing (see Table A.14, Appendix A). There are no universal standards associated with geographic information science certificates. Rather, each institution determines the course content and number of credit hours required to complete the certificate. Most geographic information science certificates are awarded for advanced theoretical and technical training. For example, the University of Texas, Dallas, offers two 15-hour graduate certificates in remote sensing: one focused on remote sensing and digital image processing and the other focused on the application of geospatial ideas and techniques to national security and intelligence. The National Center for Remote Sensing, Air, and Space Law at the University of Mississippi offers a 12-hour certificate in remote sensing, air, and space law.[4] Table A.15 (Appendix A) summarizes the course requirements for a certificate in GIS with an emphasis in remote sensing at the University of Utah.

A few certificates are awarded for taking a single course, independent of a formal academic degree. For example, the University of Twente International Institute for Geo-Information Science and Earth Observation offers certificates in a variety of geospatial topics, including remote sensing, hyperspectral remote sensing, remote sensing and digital image processing, cartography and geovisualization, and principles of GIS.[5]

Interdisciplinary Programs

Although there is an enduring need for deep expertise in the various scientific disciplines, demand is growing for graduates who can think and work across disciplinary boundaries to solve large and complex problems of importance to science and society (NRC, 1994, 1995; Nerad and Cerny, 1999; Nyquist et al., 1999; Nyquist, 2000; Golde and Dore, 2001; Pallas, 2001; Lélé and Norgaard, 2004). An interdisciplinary perspective can be helpful for dealing with a variety of geospatial intelligence issues, such as those that concern coupled human-environmental systems (e.g., the national security implications of climate change, economic globalization, poverty, and transborder migration). Interdisciplinary programs commonly begin at the initiative of faculty, although a few agency programs are available to support their development.

---

[3] See <http://www.worldcampus.psu.edu/MasterinGIS.shtml>.

[4] See <http://www.spacelaw.olemiss.edu/academics/certificate.html>.

[5] See <http://www.itc.nl/Pub/study/Programmes/Certificate>.

Integrative Graduate Education and Research Traineeship (IGERT) Program. The National Science Foundation (NSF) established the IGERT program to produce Ph.D. scientists and engineers with interdisciplinary training, discipline knowledge, and technical, professional, and personal skills (e.g., communication, ethics, teamwork, leadership) that are useful to both academic and nonacademic careers.[6] IGERT grants are awarded competitively to university faculty, and projects can be funded for as long as 10 years. The projects are organized around an interdisciplinary science theme but include opportunities for hands-on experience, work in other countries, and professional development (e.g., internships) that complement academic preparation.

A number of IGERTs have touched on the core or emerging areas discussed in this report. For example, the Sensor Science, Engineering, and Informatics project at the University of Maine is examining all aspects of sensor systems, from the science and engineering of new materials to the interpretation of sensor data. The objective is to use knowledge from sensor-generated data to drive development of sensor systems and advances in sensor materials and devices, and vice versa. A recent IGERT with an explicit geospatial focus was the Integrative Geographic Information Science Traineeship project at the State University of New York, Buffalo. The project facilitated interdisciplinary research in geographic information science, environmental science (e.g., integration of spatial databases with regional models to forecast environmental changes), and social science (e.g., integration of spatial analysis and spatial statistics with GIS to detect crime or disease hot spots). As part of the project, the IGERT team provided a rapid, large-scale damage assessment following the 2010 Haiti earthquake.[7]

A 2006 assessment found that the IGERT program has had a measurable impact on students, faculty, and institutions (Carney et al., 2006). Students in IGERT programs reported feeling well grounded in their discipline but better prepared to work in multidisciplinary teams and to communicate with people in other disciplines compared to their non-IGERT peers. In addition, IGERT faculty increased their interdisciplinary work, leading to new collaborations, research

ideas, and courses, and, in some cases, to stronger institutional support for interdisciplinary approaches. For example, some IGERTs have become self-sustaining Ph.D. programs after NSF support ended (Box 6.1).

University of Southern California (USC) Joint Games Program. Games have moved beyond simple entertainment to become tools that support a variety of applications, including military recruitment and training (e.g., America's Army game;[8] NRC, 1997) and training in human geography. For example, "Sudan: Darfur is Dying" is a narrative video game that simulates the experience of 2.5 million refugees in the Darfur region (Figure 6.2). Players deal with threats to the survival of their refugee camp, such as possible attack by Janjaweed militias. They can also learn more about the genocide, human rights, and the humanitarian crisis in Darfur.

Games are inherently multidisciplinary, requiring designers, engineers, and artists to come together to design the gameplay and visuals and to program the software. Games programs at universities also have an interdisciplinary element. A good example is the USC joint Games Program, which was created by Michael Zyda, a member of this committee. The Department of Computer Science at USC offers a bachelor's program in computer science (games) and a master's program in computer science (game development; Zyda, 2009). The master's program requires computer science students to take three game design courses in the School of Cinematic Arts, which are aimed at getting designers and engineers used to working together and learning each other's strengths.

Students interested in joint game building spend their last year building games in large teams of designers, computer scientists, and artists from a wide range of departments (Figure 6.3). The joint Games Program 491 course begins in the spring with a call for game designs. A panel of industry and faculty members chooses which designs will be developed, and the leads for the game designs chosen flesh out their game design and recruit their teams by the first day of fall classes. During the fall semester, the teams develop a playable prototype. In the spring semester, midcourse corrections are made

---

[6] See <http://www.nsf.gov/publications/pub_summ.jsp?WT.z_pims_id=12759&ods_key=nsf11533>.

[7] See <www.igert.org>.

[8] The America's Army game, initially built to support Army recruiting, now has more than 13 million registered players and has become the core platform behind many Army training systems. See <http://info.americasarmy.com>.

BOX 6.1
From an IGERT Project to an Interdisciplinary Ph.D. Program

In 1999, Kathleen Carley, committee member and professor at Carnegie Mellon University, received an IGERT award to study social complexity and change using computational analysis of social and organizational data. The primary methods used in the project were network analysis, agent-based modeling, and statistical models of dynamic systems, and students took courses in network analysis, computer simulation, statistics, algorithms or machine learning, and organization or policy science. Students could come from any department in the university, and the course of study was overseen by advisors from the home department, a computer science department, and a social science department. Over the 10-year lifespan of the project, 18 students received some support from the program and 10 others became affiliates.

Shortly after the IGERT project began, several other educational initiatives that blended computer science and social, organizational, or policy science were started at Carnegie Mellon University. Because these programs were spread over many departments and colleges, mentoring the grow-ing number of students became increasingly difficult. By 2001, several faculty with interdisciplinary interests in computer science and the social and organizational sciences had moved their appointments to the Institute for Software Research, a new department in the College of Computer Science. These faculty banded together to form the Computation and Organization Science (COS) Ph.D. program, which became a standing program in 2004.

The COS curriculum was designed after the IGERT curriculum, but was expanded to include a policy component. The COS program is aimed at producing new Ph.D.s capable of (1) assessing the social or policy impact of new computational technologies, such as crowdsourcing technologies for disaster response; and (2) designing, developing, and testing new computational technologies that will affect humans at the societal, cultural, or policy level, such as new cell-phone applications that track and share the movements of individuals. Students are taught the basics of social network analysis and the advanced methods integral to dynamic network analysis, and the program of study is tailored through electives that can be taken in any of the colleges at Carnegie Mellon University. Special attention is placed on geo-enabled network analytics. Where the goal is to track the region of influence of actors of interest or to identify how to disrupt terror or piracy networks in a region, the combination of social and spatial information and the use of unified tools is critical. Through a combination of project-based courses and research, the students acquire knowledge, skills, and practical experience needed to contribute to advances in these areas.

Today the COS program has 27 Ph.D. students and 18 alumni who are employed in both industry and academia. Students trained in networks, and in particular those with a strong computational or geo-network background, are in high demand. However, the number of qualified students applying to conduct research in this area far exceeds the available fellowships, research grants, and industry support for master's programs.

FIGURE 6.2 Screenshot of a Sudan game intervention: spreading beliefs to reduce levels of intertribal hostility. SOURCE: Courtesy of the USC GamePipe Laboratory.

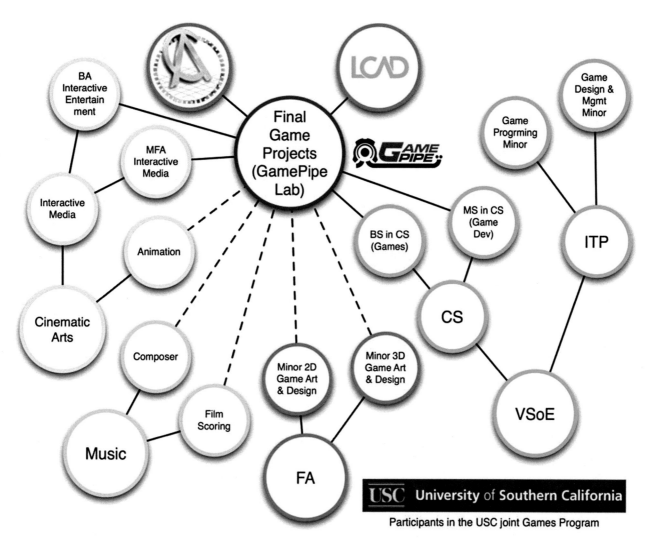

FIGURE 6.3 Participants in the joint USC Games Program, including several USC departments, Atlantic College (Puerto Rico), and the Laguna College of Art and Design (LCAD) Game Art Program. The solid lines indicate strong participation; dashed lines indicate some participation. NOTE: CS = Computer Science Department; FA = Fine Arts Program; ITP = Institute of Creative Technologies; VSoE = Viterbi School of Engineering.

and the teams continue developing their games. At the end of the semester, each team demonstrates a polished product with 2–4 hours of gameplay to fellow students and industry executives.

Building the USC joint Games Program required time, effort, and courage. The question of how and when to support a new area is difficult for most universities. Moreover, the university department structure works against multi- and interdisciplinary programs, and issues about where to establish the departmental home, where to manage the budget, how to evaluate the faculty, and how to obtain enough space had to be resolved. However, once established, the program has proven a success. The joint Games Program was rated the top game design program by the Princeton Review in March 2011,[9] and it is not unusual for the student teams to be recruited by the gaming industry.

Summer Programs

Advanced techniques and methods are commonly taught in university summer courses. The largest U.S. summer program in GIS and spatial analysis is offered by the Interuniversity Consortium for Political and Social Research (ICPSR), which is based at the Uni-

___
[9] See <http://www.hollywoodreporter.com/news/usc-named-top-school-video-162975>.

versity of Michigan. ICPSR offers 5-day workshops dealing with introductory spatial regression analysis and spatial econometrics. Its counterpart in the United Kingdom is the University of Essex Summer School in Social Science Data Analysis, which offers a 2-week intensive course on spatial econometrics. The Spatial Econometrics Association runs a 4-week Spatial Econometrics Advanced Institute every summer in Rome.

Summer courses in network analysis are offered by ICPSR and Carnegie Mellon University's Center for Computational Analysis of Social and Organizational Systems (CASOS) and COS Ph.D. program. The CASOS summer institute provides continual reeducation on geo-enabled network analysis and dynamical analysis. Courses focus on new tools for analyzing network data, which are evolving rapidly; changes in technologies for data collection, visualization, and forecasting that influence the analysis of network data; and advanced analyses that have a spatial, temporal, or large-scale data component. All COS Ph.D. students have the opportunity to take this program and teach in it, and the new advances that they develop are presented as part of an associated symposium.

GOVERNMENT

Government training programs are established to meet the specialized needs of federal agencies. Military colleges offer an opportunity to train students in relevant disciplines and techniques within the context of national security and defense. Training programs offered by federal agencies provide a means for employees to update or augment necessary skills. Such training is important for geospatial intelligence, given the rapid advances in geospatial technologies and the need for operational knowledge, experience, and tactics, which cannot be fully taught in academic settings.

NGA Vector Study Program

The Vector Study Program, originally called the Long Term Full Time Training (LTFTT) program, was established more than 50 years ago to enable civilian employees of NGA's predecessor organizations to obtain university training in selected areas while re-

ceiving full salary and benefits. NGA identifies critical skills needed, interested employees apply for specific programs, and universities accept applicants through the normal selection process. The program pays for tuition and all course-related expenses for classes in bachelor's, master's, and doctorate programs. The master's program is funded for 1 year (three semesters) and the doctorate program is funded for 2 years (six semesters). Obtaining a degree is not required, and indeed most advanced degrees can barely be completed within these time periods.

About 70 universities and colleges currently participate in the Vector Study Program. One of the longest running programs is the graduate program in photogrammetry at Purdue University, which was established in the mid-1960s primarily to train employees of federal agencies, including the Defense Mapping Agency, a predecessor of NGA. The program offers master's degrees, both thesis (research oriented) and nonthesis options, and doctorate degrees. Most trainees elected the master's nonthesis option, which offers a large number of courses (10–12). The few who chose the thesis option usually worked on topics related to their agency's activities, seeking solutions to problems encountered in production. Ph.D. trainees commonly worked on state-of-the-art problems of interest to their agency.

Topics covered in Purdue's graduate program include the fundamentals of photogrammetry, sensor (passive, active, motion-imagery) modeling, platforms (space, airborne, unmanned), lidar, information extraction (targeting, image products such as ortho-imagery, digital elevation models, map features), and photogrammetric data as input to Computer Aided Design, GIS, and virtual reality databases. Other allied courses in the program include analytical fundamentals for geomatics, adjustment of geospatial observations, coordinate systems and conformal mapping, advanced geospatial estimation, modeling, and exploitation of multi- and hyperspectral remote sensing systems. Students who stay beyond the first year enroll in courses in related areas, such as orbit mechanics (for integration into photogrammetric modeling) and digital image processing and analysis (learning the approaches to image feature correspondence). They delve deeply into advanced photogrammetric problems, such as image registration (image to image and image to reference data) and the complex task of image and data fusion. Additional courses include GIS, remote sensing, geodesy, statistics, navigation, radar, numerical analysis, and databases.

The nonthesis master's degree at Purdue has trained hundreds of employees of NGA and its precursor agencies in its 50 years of existence. In addition, several individuals received doctorates and went on to become technical leaders within the agency.

NOAA National Weather Service

The National Weather Service (NWS) issues weather forecasts and warnings for the protection of life and property. Weather forecasters fuse results from dynamic model projections, weather images from radars or satellites, ground observations, spotter reports, and their own understanding of storm evolution and atmospheric dynamics to determine the timing, duration, and spatial extent of the warning. The need to discern actionable weather intelligence under tight time pressure mimics what is needed for geospatial intelligence. Moreover, weather information and technologies are innately geospatial, and the skills needed to prepare a weather forecast and issue a warning (e.g., information synthesis, situation awareness) are similar to those needed by NGA scientists and analysts. Thus, NWS training may be a useful model for NGA training, particularly in the areas of forecasts, GEOINT fusion, and geospatial analysis.

The NWS operates three training facilities, including the Warning Decision Training Branch in Norman, Oklahoma. The warning decision facility provides in-residence training workshops and distance learning modules on weather radar operations, particularly the integrated data environment, warning methodology, situation awareness, and decision making.[10] In the advanced warning operation course, students are taught the theoretical underpinning of situation awareness, then are given exercises, developed from past weather events, to simulate severe weather operations. These exercises allow students to practice advanced warning concepts in an operational context under different severe weather scenarios. The curriculum is adjusted each year to respond to changing NWS strategic goals

and operational requirements and to incorporate the results of evaluations of previous training (NWS, 2008).

Military Colleges

Some U.S. military colleges and universities offer classes or degrees in geospatial topics, often in the context of intelligence or national defense. The U.S. Military Academy offers a bachelor's degree in geography and environmental engineering, with courses related to environmental geography, human geography, cartography, and geographic information science. The U.S. Air Force Academy offers a bachelor's degree in geography, with courses in remote sensing, political geography, human geography, cartography, and spatial analysis. The other academies provide some geospatial training within other programs (e.g., ocean science, strategic intelligence and homeland security). At the graduate level, both the Naval Postgraduate School and the Air Force Institute of Technology offer geography courses and courses that use GIS and human geography technologies, although neither has a degree program in this area. Because students generally go into service following graduation, their formal training can be outdated by the time they enter the workforce. However, their experience within the services commonly refines and updates their applied geospatial skills.

PROFESSIONAL SOCIETIES AND NONGOVERNMENTAL ORGANIZATIONS

Short courses and conference workshops offered by professional societies (e.g., Table 6.1) and other nongovernmental organizations provide immersive, short-term training on a variety of geospatial topics of potential interest to NGA. For example, the Association of American Geographers (AAG) hosts workshops on software applications for geospatial analysis. Geospatial technology workshops are led by credentialed experts and typically run a half or full day as part of the annual conference program. The American Society for Photogrammetry and Remote Sensing (ASPRS) offers numerous remote sensing workshops and webinars each year. The workshops and webinars in 2011, for example, focused on hyperspectral and lidar instruments and on accuracy assessment. Some of these

---

[10] See <http://www.wdtb.noaa.gov/courses/>.

TABLE 6.1 Major Professional Societies Offering Training in Core or Emerging Areas

Professional Society	Mission	Subject Areas
American Geophysical Union (AGU) <http://www.agu.org/>	Promote discovery in Earth and space science for the benefit of humanity	Geodesy, geophysics, remote sensing
American Planning Association <http://www.planning.org/>	Advocate excellence in community planning, promote education, and provide tools and support to meet the challenges of growth and change	Human geography
Association of American Geographers (AAG) <www.aag.org>	Advocate the understanding of the world's geography, including human and physical systems and the use of geographic techniques	Cartography, GIS and geospatial analysis, human geography, remote sensing
American Society for Photogrammetry and Remote Sensing (ASPRS) <http://www.asprs.org/>	Promote the ethical application of active and passive sensors, photogrammetry, remote sensing, and geospatial technologies; and advance the understanding of geospatial and related sciences	Cartography, GIS, photogrammetry, remote sensing
Association for Computing Machinery Special Interest Group (ACM SIG) spatial <www.sigspatial.org/>	Address issues related to the acquisition, management, and processing of spatially related information	Cartography, GIS, remote sensing
Cartography and Geographic Information Society (CaGIS) <http://cartogis.org/>	Connect professionals who work in cartography and geographic information science, both nationally and internationally	Cartography, GIS, photogrammetry, remote sensing
IEEE Geoscience and Remote Sensing Society (GRSS) <http://www.grss-ieee.org/>	Advance science and technology in geoscience, remote sensing and related fields	Geodesy, photogrammetry, remote sensing
Institute of Navigation (ION) <http://ion.org/>	Advance the art and science of positioning, navigation, and timing	Geodesy, remote sensing, cartography
International Society for Photogrammetry and Remote Sensing (ISPRS) <http://www.isprs.org/>	A society of national societies and organizations specializing in photogrammetry and remote sensing	Photogrammetry, remote sensing
Society of Photographic Instrumentation Engineers (SPIE) <http://spie.org/>	Advance light-based technologies	Geodesy, photogrammetry, remote sensing
University Consortium for Geographic Information Science (UCGIS) <http://www.ucgis.org>	Serve as a voice for the geographic information science research community, foster multidisciplinary research and education, and promote the use of geographic information science and geographic analysis for the benefit of society	Cartography, GIS, remote sensing
Urban and Regional Information Systems Society (URISA) <http://urisa.org/>	Use GIS and other information technologies to solve challenges in state, regional, and local government agencies	Cartography, GIS, human geography, remote sensing

workshops and webinars provide opportunities for participants to obtain formal continuing education credit needed for employment, certification, or recertification.

A number of conferences include short courses ranging from several hours to several days focusing on particular software programs, advanced methods in GIS or geospatial analysis, or other specialized topics. For example, the Cartography and Geographic Information Society (CaGIS) offers cartography and GIS-related short courses, such as the 1-week course

on spatial statistics with open-source software offered during the GeoStat 2011 Conference. The Institute of Navigation (ION) offers geodesy-related short courses for professional development in positioning, navigation, and timing. For example, the 2011 ION annual GNSS international technical meeting included 16 courses on GPS/GNSS from beginner to advanced levels. Short training courses on social networks are held regularly at the main International Network for Social Network Analysis meeting and sometimes at meetings of the

Academy of Management, the American Sociological Association, and the Institute for Operations Research and the Management Sciences. These training sessions have helped spread new capabilities and also contributed to growth in the field.

INDUSTRY

Several companies provide training in using the tools they have developed, some of which are being used by NGA. Among the most notable is the Environmental Research Systems Institute (ESRI), a company specializing in GIS software and geodatabase management applications. ESRI hosts an annual user's conference, which is a forum for training individuals to use ESRI products. Courses, which are either taught by instructors or are self-paced, focus on GIS technology skills and practices to accomplish GIS workflows on the desktop, server, and mobile web.[11] BAE Systems offers courses on its photogrammetric products and workstations with SOCET GXP software in several cities.[12] Intergraph Corp. teaches 1- to 4-day classes in geospatial technology, video systems, and its geo-media software on request in Madison, Alabama.[13] In addition, some manufacturers of GPS/GNSS receivers offer training courses to their customers. Navtech GPS currently offers 12 courses, some of which deal with the geodetic aspects of GPS/GNSS, at both public venue and corporate facilities.[14] Most of these courses are based on proprietary systems, although training based on open-source software and tools is beginning to emerge (e.g., open-source GIS boot camp offered by Geospatial Training Services LLC).[15]

SUMMARY AND CONCLUSIONS

The third task of the committee was to describe training programs for geospatial intelligence disciplines and analytical skills. Training programs relevant to geospatial intelligence abound in universities, govern-

ment agencies, professional societies, and industry. Universities provide the widest variety of training programs, ranging from comprehensive degree programs that cater to individuals preparing for a career in geospatial science or technology to specialized programs and certificates aimed at expanding or updating the knowledge and skills of professionals. Although a few universities offer degrees in interdisciplinary fields, such as those related to the emerging areas, such programs remain difficult to create and sustain. Training offered by government agencies, professional societies, and private companies focuses on education and professional development of individuals already in the workforce and generally takes the form of workshops, short courses, or classes on particular tools or techniques. Distance learning is increasingly an option for all types of training.

The training programs chosen to address Task 3 have a long record of accomplishment, a critical mass of high-caliber instructors, a substantial number of students, and/or provide an opportunity to solve problems in a real-world context. The example programs and their relevance to NGA are as follows:

- A typical undergraduate curriculum, which provides a foundation of geospatial knowledge and skills (e.g., University of Colorado's Department of Geography).
- Master's of science programs, which emphasize solutions to real-world problems (e.g., George Mason University's master's program in geographic and cartographic sciences).
- Professional science master's program aimed at producing graduates with a mix of scientific expertise and practical workforce skills (e.g., North Carolina State University's professional science master's program in geospatial information science and technology).
- Degrees or classes in geospatial topics in the context of intelligence or national defense (e.g., military colleges).
- Interdisciplinary programs, which are useful for dealing with complex geospatial intelligence issues (e.g., Carnegie Mellon University's Computational and Organization Science program; University of Southern California's Joint Games Program).
- A mature distance learning program (e.g., Pennsylvania State University's GIS curriculum).

---

[11] See <http://www.esri.com>.

[12] See <http://www.socetgxp.com/content/events/training-courses>.

[13] See <http://www.erdas.com/service/training/training.aspx>.

[14] See <www.navtechgps.com>.

[15] See <http://www.geospatialtraining.com/index.php?option=com_catalog&view=node&id=71%3Aopen-source-gis-bootcamp&Itemid=108>.

• Certificates acknowledging special training in selected subject matter (e.g., University of Utah graduate certificate in GIS with an emphasis in remote sensing).

• Summer programs, short courses, and workshops aimed at teaching advanced techniques and methods (e.g., Interuniversity Consortium for Political and Social Research's summer program in GIS and spatial analysis; ION's short courses in positioning, navigation, and timing; ESRI's annual user's conference, with short courses on GIS technology and skills).

• Programs that send employees for training at universities while receiving full salary and benefits (e.g., NGA's Vector Study Program).

• Operational training in issuing forecasts and warnings (e.g., National Weather Service's Warning Decision Training Branch).

7

Building Knowledge and Skills

Chapter 6 described the various types of university, government, professional society, and industry programs that currently offer some type of education or training in geospatial disciplines, methods, or technology. Although not designed for the National Geospatial-Intelligence Agency (NGA), many of these programs could be used by NGA to obtain some useful knowledge or skills. NGA could also take steps to build the specialized expertise it needs in the future.

NGA has a number of useful programs for building knowledge and skills in specific areas or expanding the diversity of the pool of applicants. These include programs to train current employees (e.g., NGA College, Vector Study Program), grants to academic institutions and consortia to support NGA-relevant research and education, and scholarships and internships to support students interested in pursuing a career in geospatial intelligence (Box 7.1). This chapter focuses on other actions NGA can take to build the specialized knowledge and skills it needs to ensure an adequate U.S. supply of geospatial intelligence experts, including the emerging areas, over the next 20 years (Task 4). The objective was to provide a menu of choices of varying scope, not to identify priorities. The ideas are organized into three categories: building new knowledge in the core and emerging areas, strengthening existing training programs, and enhancing recruitment efforts.

BUILDING THE CORE AND EMERGING AREAS

Most of the education and training in core and emerging areas takes place in universities. NGA supports some of these efforts by funding research projects and students (Box 7.1). Funding could also be directed toward building university programs, curricula, and academic support infrastructure to help develop fields of interest to NGA, as described below. Partnerships with other agencies (e.g., National Science Foundation [NSF], National Aeronautics and Space Administration [NASA], National Oceanic and Atmospheric Administration) on mutual topics of interest would stretch NGA's research dollars and help sustain initiatives long enough to ensure a sufficient supply of experts over the next 20 years.

University Affiliated Research Centers

University Affiliated Research Centers (UARCs) are government research centers attached to universities at the forefront of a specific research area. The Department of Defense (DOD) began establishing UARCs in 1996 to help maintain core engineering and technology capabilities. The initial set of centers is still operating and more have been added, attesting to their usefulness to DOD. There are currently 13 DOD UARCs, none of which are focused on geospatial technology or applications. An NGA UARC could support geospatial intelligence areas that would not otherwise exist in the university; foster ongoing collaboration among research faculty, Ph.D. students, and NGA staff; and maintain

---

BOX 7.1
NGA Grants, Scholarships, and Internships

NGA's Academic Research Program awards research grants to universities to support basic research of interest to the agency, to fill gaps in imagery or geospatial science and technology, and/or to develop associated education and training programs.[a] NGA also offers scholarships and paid internships to college students.[b] Programs include the following:

NGA University Research Initiatives—support research in geospatial intelligence disciplines at U.S. colleges and universities that carry out science and engineering research and/or related education.

Historically Black Colleges and Universities/Minority Institutions Grants—support educational research to develop and enrich geospatial research and teaching environments at historically black colleges or universities and minority serving institutions.

NGA Outstanding New Scientific and Technical Innovative Researcher Program—support innovative NGA-relevant research by faculty members who have held their doctorate degrees for less than 5 years.

NGA Research Collaboration Forums—encourage collaboration among educational institutions that carry out science and engineering research to advance scientific breakthroughs or improve understanding of research areas of interest to NGA.

Service Academy Education—support basic research and education-related research activities in geospatial sciences at the U.S. service academies.

Visiting Scientist Program—place visiting academic researchers in NGA facilities.

NGA Student Employment Program—provide summer internships to undergraduate and graduate students to give them real work experience and prepare them for future employment with NGA.

NGA Stokes Scholarship Program—provide college undergraduates who have demonstrated financial need and interest in an NGA career with tuition assistance, challenging summer work, and a guaranteed position in their field of study upon graduation.

Science, Mathematics, and Research for Transformation (SMART) Program—provide a full scholarship, stipend for living expenses, and employment in the federal government upon completion of a degree in science, technology, engineering, or mathematics. NGA is a participating placement site for the scholars.

---

[a] See <https://www1.nga.mil/PARTNERS/RESEARCHANDGRANTS/Pages/AcademicResearchProgram.aspx>.
[b] See <https://www1.nga.mil/Careers/StudentOpp/Pages/default.aspx>.

---

a long-term research and development focus on areas critical to NGA.

A new UARC is usually initiated at a high level within an agency. For example, the University of Southern California's Institute for Creative Technologies was initiated by the chief scientist of the U.S. Army at the time, A. Michael Andrews, II, who was inspired by the NRC (1997) report *Modeling and Simulation—Linking Entertainment and Defense.* Dr. Andrews requested the chair of the NRC committee to draft a research agenda and operating plan for a UARC focused on using entertainment technologies

to develop the Army's next generation of immersive training environments. Funding for the center, initially $10 million per year for 5 years, has grown to more than $25 million per year.

Centers of Excellence

Centers of excellence are commonly established to carry out collaborative research, create tools and data sets, and build a cohort of trained individuals in subject areas outside traditional academic, business, or government departments. They are similar to UARCs, but can be situated at universities, government agencies, national laboratories, or private companies; and they can cover any topic that requires a team approach or shared facilities.

The Intelligence Community Centers of Excellence, which are partly supported by NGA, are focused on improving the representation of minorities and women in critical competencies, such as information technology, language, political science and economics, science and engineering, and threat analysis.[1] An example of centers that both generate mission-specific knowledge and train the next generation of experts are the Department of Homeland Security (DHS) centers of excellence. Established as part of the Homeland Security Act of 2002,[2] the centers are intended to enhance homeland security by generating knowledge and ideas for new technologies in a wide range of subjects. Major themes of the centers include terrorism; microbial risk; zoonotic disease; food security; preparedness; explosive-related threats; border security; maritime and remote resources; coastal areas; transportation; and command, control, and interoperability. Each center is led by a university, often in collaboration with other universities, national laboratories, nongovernmental organizations, government agencies, and private companies. A few of the centers touch on emerging areas discussed in this report. For example, the Center of Excellence in Command, Control, and Interoperability at Purdue University covers visual analytics for security applications. Many of the centers also provide education and training to students and/or professionals. For example, the National Consortium for the Study of Terrorism and Responses to Terrorism offers a graduate certificate in terrorism analysis and an undergraduate minor in terrorism studies, and the Center for Maritime, Island and Remote and Extreme Environment Security offers professional development courses in port-security sensing technologies.

Centers of excellence can be effective sources of innovation, especially those housed in private companies (e.g., Frost et al., 2002). Centers located in universities also create a culture that links research and

the government, potentially facilitating recruitment and increasing the pool of graduates with knowledge and skills needed by the sponsor agency (or agencies), depending on how long funding is sustained. Multiple years of support are commonly required to build education programs as well as to fund graduate students with dissertation or thesis topics of direct interest to an agency.

Virtual Centers

UARCs and centers of excellence have a physical home, although many participants do not work there. A virtual center may or may not have a home, but usually consists of a leader and appointed or self-selected members who work on a common goal from their own institutions. Virtual centers are easy to establish (and disestablish) and relatively inexpensive to operate, and the structure can be customized to the need. Maintaining a virtual center can be as simple as providing a web server and supporting conferencing.

Virtual centers are often created where fields are evolving rapidly and the necessary skills and knowledge for advancing them are scattered across many institutions. Indeed, virtual clearinghouses for curriculum and contact information were essential for the development of Geographic Information Systems (GIS) studies (e.g., Kemp and Goodchild, 1992) and are now being used to help develop the visual analytics field (Thomas and Cook, 2006). Virtual centers could provide several types of benefits to NGA, including providing a means to build emerging areas in universities or to facilitate collaboration among NGA offices or partner organizations.

An example of a virtual center focused on facilitating research collaborations is the Research Information Centre, which was developed jointly by Microsoft Research Connections and The British Library. The center provides management software tools, such as domain-specific project site templates, calendars, task lists, wikis, blogs, and surveys (Barga et al., 2007). Tools for automating collaboration among research groups across locations and disciplines are now under construction (Procter et al., 2011). Some of these are entire learning systems aimed at instruction; some support the building of common reference data, bodies of knowledge, and toolsets; and others are geared toward

---

[1] See <www.nsu.edu/iccae/pdf/IC-CAEGuidanceAndProcedures.pdf>. A list of centers can be found at <http://www.nsa.gov/ia/academic_outreach/nat_cae/institutions.shtml>.

[2] Public Law 107-296. See also <http://www.dhs.gov/homeland-security-centers-excellence>.

removing the impacts of physical separation on collaborative research. A study on the project found that shared access to data, tools, computational resources, and collaborators has led to faster research results and novel research directions (Carusi and Reimer, 2010).

Research Partnerships

Innovation in geospatial technology commonly comes from industry or collaborations with industry. Examples of such technologies used by NGA include ArcGIS Military Analyst, which was developed by the Environmental Systems Research Institute (ESRI), and FalconView, a PC-based mapping application developed by the Georgia Tech Research Institute. One way to nurture nonproprietary technology innovations is to develop research partnerships with private companies. Cooperative Research and Development Agreements (CRADAs) are commonly used to establish research and development partnerships between a government agency and a private company. Partnerships between universities and industry can be formed through a variety of means. For example, NSF's Industry & University Cooperative Research Program provides a means for universities and private companies to establish a center, supported primarily by industry, to collaborate on projects of mutual interest.[3] The program is intended to help build the nation's research infrastructure and to enhance the intellectual capacity of the science and engineering workforce. Private companies provide funding and technological capabilities, and universities provide cutting-edge research capabilities. Graduate students contribute to the research projects and also become familiar with industrially relevant research.

Some of the centers in the Industry & University Cooperative Research Program address topics of interest to NGA, such as remote sensing, visual analytics, and data fusion. Government agencies can become partners in the centers or use this model to build critical infrastructure and worker skills specific to their needs.

Curriculum Development

A number of federal agencies have sponsored initiatives to develop or enhance curricula in areas relevant to their mission, thereby helping to expand the supply of potential employees with the necessary training and skills. For example, NASA has sponsored several projects to develop remote sensing curricula.[4] Opportunities abound for NGA to get involved in curriculum development in emerging or other areas that suit their workforce needs. A particularly promising focus is an interdisciplinary master's degree curriculum in geospatial intelligence topics. Interdisciplinary master's programs are politically easier and less costly for universities to implement than interdisciplinary bachelor's programs. Moreover, efforts to establish such curricula would demonstrate to universities the need for interdisciplinary education. Curriculum development at the NGA College may also be fruitful. Such efforts are often inexpensive and can yield major returns.

Past experience with creating academic curriculum in emerging geospatial areas is well illustrated by the NSF's National Center for Geographic Information and Analysis (NCGIA), which was established in 1989. The proposal to create the center included development of a core curriculum in GIS. At the time, no major textbook on the subject had been written and few universities offered classes. Therefore, the initial NCGIA core curriculum, published in 1990, was targeted at university and college instructors and included lesson plans, lecture slides, and support materials. The curriculum was a success, with requests for the materials from hundreds of institutions nationally and internationally (Kemp and Goodchild, 1991). Ongoing demand led to a second version of the GIS core curriculum, this time using the web as the main creation and distribution channel. An overall design and structure was created, and leading scholars were invited to contribute content to each of the modules. Although the web version of the GIS core curriculum was overtaken by Wikis and by new textbooks and software, its model for basic classes and topics remains at the forefront of university-level GIS instruction today (Howarth and Sinton, 2011).

Academic Support Infrastructure

The academic support infrastructure for the emerging areas—professional societies, special interest

---

[3] See <http://www.nsf.gov/eng/iip/iucrc/index.jsp>.

[4] See <http://www.icrsed.org/hist.html>.

groups, journals or special issues of journals, workshops, conferences, websites, and blogs—is still in its infancy (see Chapter 3). Although such support systems will come as the fields develop in academia, NGA may be able to encourage their growth by increasing awareness of the emerging areas and their interest to NGA. Possible actions include the following:

• Funding a university scientist to edit a special issue on an emerging topic in a journal by soliciting articles from colleagues.
• Creating a blog (classified or unclassified).
• Soliciting articles by leading academics on the emerging areas for NGA's *Pathfinder* Magazine.
• Funding individuals to write wikis or maintain a clearinghouse of executable software used in research.
• Sponsoring sessions on emerging themes at key conferences.

Success could be measured by the emergence of formal academic infrastructures (e.g., journals, society interest groups) that are self-supporting or by the number of articles in the emerging areas and their citation counts.

STRENGTHENING TRAINING

NGA trains its employees and contractors primarily through the Vector Study Program and the NGA College. Actions NGA can take to strengthen training offered by these programs and other opportunities to train current employees are described below.

Vector Study Program

The Vector Study Program has produced a relatively large number of NGA employees with advanced skills and training, particularly in photogrammetry and geodesy. However, enrollments in the core areas have been declining, jeopardizing the viability of academic photogrammetry programs, and Vector Study Programs do not exist in the emerging areas. Because academic programs in many areas of interest to NGA are already in place, expanding and/or modifying the Vector Study Program would result in nearly immediate gains in staff trained in critical areas.

Although most NGA employees receive specialized training at the NGA College, class offerings in the core areas are limited compared to those offered by a top university degree program (Box 5.2). As a result, current employees are receiving less in-depth training than employees who are nearing retirement. Increasing enrollments in the Vector Study Program could forestall a loss of skill. In addition, the NGA College offers few classes in the emerging areas. Adding graduate programs in emerging areas to the Vector Study Program would produce NGA employees with new skills. About one-third of universities that participate in the Vector Study Program have departments that provide strong education and training in an emerging area (see Tables A.6–A.10 in Appendix A). These universities may be good near-term candidates for Vector Study Programs in emerging areas. Because Vector Study Programs are developed by university faculty members in collaboration with NGA, the new programs would also allow NGA to influence developments in the field.

The flexibility of the Vector Study Program could be increased by including online or distance-learning classes in the program, which would allow employees to take courses while working part-time at NGA. Once a sufficient number of online credits have been acquired, the employee could complete the degree requirements on campus. The combination of online and on-campus study could be tailored to suit the individual and/or program need. Another way to increase program flexibility is to allow both shorter and longer periods of study. The program currently specifies a number of semesters in a particular period (e.g., two semesters and a summer session in one calendar year for a nonthesis master's degree in photogrammetry). A midcareer employee could benefit significantly from even a single semester of refresher courses, advanced courses in their specialty area, or introductory courses in new or emerging areas. Other employees could benefit from a longer course of study, such as an extra year. Ph.D. programs in particular are difficult to finish in the time allowed by the program, and both online and distance learning and longer periods in residence on campus would facilitate their completion. An extension would also allow courses in multiple areas to be combined, such as language and photogrammetry. Such individuals with diverse training are needed for NGA to meet its continuously evolving responsibilities.

Reviews of the NGA College

University departments commonly rely on external reviews to obtain feedback on past performance and ideas for future directions. The reviews can be formal or informal, and they are carried out by visiting committees of independent experts on a schedule that allows ongoing course corrections. External reviews could provide similar benefits to the NGA College.[5] Understanding strengths and weaknesses in the curriculum or faculty would help NGA College administrators ensure that the curriculum remains up to date and that the teaching staff are of the highest caliber. For the review to be independent, members of the review committee would not be associated with the college, but would be familiar with its goals and curriculum. Evaluators could be drawn from universities, professional societies concerned with geospatial science and technology (e.g., Table 6.1), and/or various branches of the armed services.

Workshops at Professional Society Meetings

Many national conferences include workshops, seminars, and training courses on specific topics, which provide an opportunity to bring NGA employees up to date on new developments. Setting up workshops and seminars is usually simple, requiring only a workshop organizer, credentialed instructors, and a mechanism for promoting the activity and registering students. By careful targeting (e.g., training in emerging areas), and by expending only small amounts of funds, it should be possible to send employees to the right workshops or to encourage the development of workshops taught by academics to meet NGA workforce needs.

An example of a venue that offers opportunities for workshops and other kinds of training in the emerging areas is the annual GEOINT Community Week, which is hosted by the U.S. Geospatial Intelligence Foundation (USGIF). The conference includes workshops, such as the 2011 workshop on analytic transformation, which covered emergent technologies and analytical methods. To date, most of the workshops and classes

have been led by government and industry instructors, but supporting university faculty to conduct workshops would bring new ideas and learning to the geospatial intelligence community beyond the technology training provided by industry, as well as expose more university faculty to NGA programs.

ENHANCING RECRUITMENT

Organizations can find qualified candidates by recruiting at universities or events (e.g., job fairs, professional society meetings) and by being highly visible to the public. NGA's small size and intelligence mission minimizes its public presence. Increasing awareness of the agency and using new approaches to find candidates with desired skills could increase the number of qualified applicants for NGA positions. Previous chapters discussed where NGA could look for candidates with geospatial skills (e.g., see "Recruiting" in Chapter 5 and Tables A.1–A.10, Appendix A). Some possible mechanisms for increasing awareness of NGA for recruiting purposes are described below.

Sessions at Professional Society Meetings

Professional society meetings and conferences are a primary means for professionals to learn about breaking research developments and to showcase their own research results. Such sessions also increase community awareness of what organizations such as NGA are doing and creates opportunities for recruiting. Employees of NGA and its predecessor organizations commonly make technical presentations at professional society meetings. For example, the Institute of Navigation has been an important venue for keeping users informed about changes in the DOD World Geodetic System 1984 (WGS 84), which provides the reference coordinate system for the Global Positioning System and the reference frame upon which all geospatial intelligence and other geospatial applications are based (e.g., Swift, 1994; Malys et al., 1997; Cunningham et al., 1998; Merrigan et al., 2002; Wiley et al., 2006). Professional society meetings also played a key role in enlisting the help of the community in evaluating various components of WGS 84 (e.g., Ture, 2004). Such community interactions, especially with students, could be leveraged to help recruit new employees.

---

[5] According to a May 23, 2011, presentation made by Mark Pahls, Chief of Learning Integration at the NGA College, classes are not formally assessed and new directions are determined in consultation with a learning advisory board.

Over the past few years, NGA has been hosting technical sessions at meetings of the American Society for Photogrammetry and Remote Sensing (ASPRS). The ASPRS Defense and Intelligence Subcommittee, which is co-chaired by NGA's senior scientist for photogrammetry, has organized both classified and unclassified technical sessions. For example, the 2011 annual conference included an unclassified special session on photogrammetry and the next generation of unmanned systems, and the 2011 ASPRS Pecora 18 Symposium included an unclassified session on the human dimensions of anticipatory intelligence analysis. Such sessions increase NGA's visibility on specific technical issues. To expand overall awareness of NGA, and thus increase recruitment opportunities, NGA could encourage students to attend the unclassified sessions to get a first-hand idea of some of the technical work at NGA. The NGA-organized sessions could also include a presentation on the agency's mission and activities, similar to what is presented at graduate seminars at universities. In addition, hosting receptions following some of these sessions would provide an opportunity for students to talk informally with NGA scientists and analysts.

Social Media Site

The next generation of NGA employees will be familiar and comfortable with the use of social media for all aspects of their daily lives, including searching for jobs and internships and exchanging information. At little expense, NGA could establish a strong social media presence that links and acts as a broker for the existing recruitment information on the NGA website. Such social media sites have been created by other defense-related agencies, such as the Australian Defence Department.[6]

By granting admission to NGA interns, employees, and others, NGA could maintain a set of highly motivated and interested users, who could be instantly informed of recruitment events, news, job opportunities, the Vector Study Program, and other topics. Features such as a director's blog or postings from NGA product users could stimulate interest and provide a broader interest group for the content of *Pathfinder* magazine, with the target of increasing recruitment.

Engaging Activities for Universities

Recruitment events at colleges, universities, and meetings are often relatively passive, with students receiving printed media and posing questions. More active engagement at such events could both provide more information to potential employees and allow individuals with the right combination of reasoning skills to be identified. For example, bringing a training exercise to a university recruitment event would enable students to actively engage in intelligence-like activities. Students could be provided with a situation to solve—such as intelligence about a facility or a natural disaster—and maps or software, and then asked to prepare and justify an analysis. Alternatively, students could be presented with results of a classic intelligence outcome and asked to analyze the decisions and the required information. Interactive feedback from recruiters is likely to be far more detailed and engaging in such an environment. Recruiters would be able to observe the students while problem solving and to judge capabilities and experience, rather than deducing it from a resume. Such activities could also be offered on GIS day[7] or provide the basis for an online quiz.

Aptitude Tests

Career and employment aptitude tests use personality tests, intelligence tests, work samples, and other tests to determine the suitability or desirability of a job applicant (Stevens and Campion, 1999). Some tests correlate better with job performance than others, so employers often use more than one test to maximize predictive power (Barrick and Mount, 1991). The use of career aptitude tests for recruiting has gained traction (Droege, 1983), including for military recruitment (e.g., Getkate et al., 1992). Internet and online submission are increasingly common for these tests.

A significant amount of research has been carried out on aptitude tests. By using custom design and existing tests, an NGA workforce targeted test could be assembled or developed relatively easily. Should such a test prove useful to NGA, it could be used in two ways: (1) as part of the diagnostic and training stage for new NGA employees and (2) as an online tool to

---

[6] See <http://www.defence.gov.au/social>.

[7] A global event to showcase real-world applications of GIS. See <http://www.gisday.com/>.

assist in recruiting. NGA could also take advantage of generic aptitude tests administered by various testing services. Individuals scoring highly on skills or native ability suited to spatial reasoning, geography, or image interpretation could easily be referred by the testing services to NGA as possible recruits.

SUMMARY AND CONCLUSIONS

The fourth task of the committee was to suggest ways to build the necessary knowledge and skills to ensure an adequate U.S. supply of geospatial intelligence experts over the next 20 years. To address the task, the committee identified a menu of NGA actions of varying scope and complexity, including the following:

• Establish research centers (UARCs, centers of excellence, virtual centers) to gather experts from different fields and/or organizations to work on issues critical to NGA.

• Establish research partnerships between private companies and universities and/or government agencies to support technological innovation.

• Sponsor university efforts to develop core curricula and academic support infrastructure (e.g., journals, conferences) needed to advance the emerging areas.

• Expand the Vector Study Program to enhance employee skills in core areas and add new skills in emerging areas.

• Institute periodic external reviews of the NGA College to ensure the quality of the curriculum and instructors.

• Send employees to short courses at professional society meetings and fund university professors to develop short courses in areas of interest to NGA.

• Increase the agency's visibility to potential job applicants by organizing sessions at professional conferences and establishing a social media site with career information.

• Seek qualified candidates by using career aptitude tests or by engaging students in interesting problem-solving exercises at recruiting events.

The examples above illustrate the variety of mechanisms that can be used to ensure the future availability of geospatial intelligence expertise. Some mechanisms would build expertise in the long term (e.g., UARCs, research partnerships with industry, curriculum development, academic support infrastructure), while others could provide more immediate gains (e.g., Vector Study Program expansion, virtual centers, professional society workshops and short courses, recruitment efforts). Most mechanisms would be relatively inexpensive to implement (e.g., virtual centers, curriculum development, recruiting efforts), while some could require substantial investment, depending on size and scope (e.g., UARCs, Vector Study Program expansion, centers of excellence). The need is greatest for the emerging areas, which have the potential to improve geospatial intelligence, but which currently produce few graduates and which lack the academic infrastructure to develop quickly. However, these mechanisms could also be used to build other areas of interest to NGA, such as core areas for which the pool of qualified applicants is small and shrinking (cartography, photogrammetry). Getting involved with education and training programs would also provide opportunities for NGA to influence the development of fields it relies on to carry out its mission.

References

Adams, J., K. Faust, and G.S. Lovasi, eds., 2012, Capturing context: Integrating spatial and social network analyses, *Social Networks*, **34** (special issue), 158 pp.

Allemag, D., and J. Hendler, 2011, *Semantic Web for the Working Ontologist*, Morgan Kauffman, Burlington, MA, 330 pp.

Anderson, B., 2010, Preemption, precaution, preparedness: Anticipatory action and future geographies, *Progress in Human Geography*, **34**, 777-798.

Andrienko, G.L., N.V. Andrienko, J. Dykes, M.-J. Kraak, and H. Schumann, 2010, GeoVA(t)—Geospatial Visual Analytics: Focus on Time, *International Journal of Geographical Information Science*, **24**, 1453-1457.

Anselin, L., 2005, Spatial statistical modeling in a GIS environment, in *GIS, Spatial Analysis and Modeling*, D. Maguire, M. Batty, and M. Goodchild, eds., ESRI Press, Redlands, CA, pp. 93-111.

Anselin, L., and A. Getis, 1992, Spatial statistical analysis and Geographic Information Systems, *The Annals of Regional Science*, **26**, 19-33.

Anselin, L., I. Syabri, and Y. Kho, 2006, Geoda, an introduction to spatial data analysis, *Geographical Analysis*, **38**, 5-22.

Antoniou, G., and F. Harmelen, 2004, *A Semantic Web Primer*, MIT Press, Cambridge, MA, 258 pp.

Barga, R.S., S. Andrews, and S. Parastatidis, 2007, A Virtual Research Environment (VRE) for bioscience researchers, in *ADVCOMP, 2007, International Conference on Advanced Engineering Computing and Applications in Sciences*, IEEE Computer Society, Los Alamitos, CA, pp. 31-38.

Barrick, M.R., and M.K. Mount, 1991, The big five personality dimensions and job performance: A meta-analysis, *Personnel Psychology*, **44**, 1-26.

Batini, C., M. Lenzerini, and S. Navathe, 1986, A comparative analysis of methodologies for database schema integration, *ACM Computing Surveys*, **18**, 323-364.

Berry, B. and D. Marble, 1968, *Spatial Analysis: A Reader in Statistical Geography*, Prentice-Hall, Englewood Cliffs, NJ, 512 pp.

Bertin, J., 1967, *Semiology of Graphics: Diagrams, Networks, Maps*, Gauthier Villars, Paris, English edition translated by W. Berg, 1987, University of Wisconsin Press, Madison, WI, 415 pp.

Bertolucci, J., 2012, Data scientists: Meet big data's top guns, *Information Week*, August 31, 2012, available at <http://www.informationweek.com/big-data/news/big-data-analytics/240006580/data-scientists-meet-big-datas-top-guns?cid=nl_IW_daily_2012-08-31_html&elq=45f156ad046144e5969b9d0fd08d29e8>.

Bivand, R., E. Pebesma, and V. Gomez-Rubio, 2008, *Applied Spatial Data Analysis with R*, Springer, New York, 378 pp.

Blachut, T.J., and R. Burkhardt, 1989, *Historical Development of Photogrammetric Methods and Instruments*, American Society for Photogrammetry and Remote Sensing, Bethesda, MD, 157 pp.

Boretos, G.P., 2011, IS model: A general model of forecasting and its applications in science and the economy, *Technological Forecasting & Social Change*, **78**, 1016-1028.

Bothos, E., D. Apostolou and G. Mentzas, 2010, Using social media to predict future events with agent-based markets, *IEEE Intelligent Systems*, **25**, 50-58.

Brealey, R., and S. Myers, 1996, *Principles of Corporate Finance*, 5th ed., McGraw-Hill, 998 pp. plus appendixes.

Cameron, K.S., and R.E. Quinn, 2006, *Diagnosing and Changing Organizational Culture: Based on the Competing Values Framework*, Jossey-Bass Business and Management Series, John Wiley & Sons, 242 pp.

Card, S., J. Mackinlay, and B. Schneiderman, 1999, *Readings in Information Visualization: Using Vision to Think*, Academic Press, San Diego, CA, 686 pp.

Carley, K.M., 2000, Organizational adaptation in volatile environments, in *Computational Modeling in Organizational Behavior: The Third Scientific Discipline*, C.L. Hulin and D.R. Ilgen, eds., American Psychological Association, Washington, D.C., pp. 241-268.

Carney, J., D. Chawla, A. Wiley, and D. Young, 2006, *Evaluation of the Initial Impacts of the National Science Foundation's Integrative Graduate Education and Research Tranineeship Program: Final Report Prepared for National Science Foundation*, Abt Associates Inc., Bethesda, MD, 34 pp., available at <http://www.nsf.gov/pubs/2006/nsf0617/nsf0617.pdf>.

Carusi, A., and T. Reimer, 2010, *Virtual Research Environment Collaborative Landscape Study*, 106 pp., available at <http://www.jisc.ac.uk/media/documents/publications/vrelandscapereport.pdf>.

CBO (Congressional Budget Office), 2012, *Comparing the Compensation of Federal and Private-Sector Employees*, Publication 4403, CBO, Washington, D.C., 16 pp.

Cheng, J., 2009, Arterial, crowdsourced traffic info comes to Google Maps, *ArsTechnica*, August 25, 2009, available at <http://arstechnica.com/web/news/2009/08/arterial-crowdsourced-traffic-info-comes-to-google-maps.ars>.

Clarke, K.C., 2009, Geospatial intelligence, in *International Encyclopedia of Human Geography*, Vol. 4, R. Kitchin and N. Thrift, eds., Elsevier, Oxford, pp. 466-467.

Clarke, K.C., 2013a, Mapping by the US intelligence agencies, in *Cartography in the Twentieth Century: The History of Cartography*, University of Chicago Press, Chicago, in press.

Clarke, K.C., 2013b, *Maps and Web Mapping*, Pearson/Prentice Hall. Upper Saddle River, NJ, in press.

Clements, M.P., and D.F. Hendry, 1999, *Forecasting Non-stationary Economic Time Series*, Zeuthen Lecture Book Series, MIT Press, Cambridge, MA, 392 pp.

Cloud, J., 2000, Crossing the Olentangy River: The figure of the Earth and the military-industrial-academic-complex, 1947-1972, in *Studies in History and Philosophy of Modern Physics*, N. Oreskes and J. Fleming, eds., Special Issue, *Perspectives on Geophysics*, September, pp. 371-404.

Collier, P., 1994, Innovative military mapping using aerial photography in the First World War: Sinai, Palestine and Mesopotamia 1914–1919, *Cartographic Journal*, **31**, 100-105.

Colwell, R.N., ed., 1983, *Manual of Remote Sensing*, 2nd ed., American Society for Photogrammetry and Remote Sensing, Falls Church, Virginia, 2,240 pp.

Cressie, N., and C. Winkle, 2011, *Statistics for Spatio-Temporal Data*, Wiley Series in Probability and Statistics, Hoboken, NJ, 624 pp.

Cunningham, J.P., E.R. Swift, and F. Mueller, 1998, Improvement of the NIMA precise orbit and clock estimates, in *Proceedings of ION GPS-98*, 11th International Technical Meeting of the Satellite Division of the Institute of Navigation, Nashville, TN, September 1998, Institute of Navigation, Washington, D.C., pp. 1587-1596.

Day, D.A., J.M. Logsdon, and B. Latell, eds., 1998, *Eye in the Sky: The Story of the Corona Spy Satellites*, Smithsonian Institution Press, Washington, D.C., 303 pp.

Delin, K.A., 2005, Sensor webs in the wild, in *Wireless Sensor Networks: A Systems Perspective*, N. Bulusu and S. Jha, eds., Artech House, Boston, available at <http://www.sensorwaresystems.com/historical/resources/DelinSensorWebsInTheWildChapter2005.pdf>.

Delin, K.A., and S.P. Jackson, 2001, The sensor web: A new instrument concept, in *SPIE's Symposium on Integrated Optics*, January, 20-26, San Jose, CA, 9 pp., available at <http://www.sensorwaresystems.com/historical/resources/sensorweb-concept.pdf>.

Delin, K.A. and E. Small, 2009, The sensor web: Advanced technology for situational awareness, in *Wiley Handbook of Science and Technology for Homeland Security*, John Wiley & Sons, New York, 14 pp., available at <http://sensorwaresystems.com/historical/resources/SensorWareSystems-SituationalAwareness-Handbook.pdf>.

de Smith, M., M.F. Goodchild, and P. Longley, 2010, *Geospatial Analysis: A Comprehensive Guide to Principles, Techniques and Software Tools*, 3rd ed., Matador, Leicester, UK, 394 pp.

DiBiase, D., M. DeMers, A. Johnson, K. Kemp, A.T. Luck, B. Plewe, and E. Wentz, 2006, *Geographic Information Science & Technology Body of Knowledge*, Association of American Geographers, Washington, D.C., 162 pp.

DiBiase, D., T. Corbin, T. Fox, J. Francica, K. Green, J. Jackson, G. Jeffress, B. Jones, B. Jones, J. Mennis, K. Schuckman, C. Smith, and J. Van Sickle, 2010, The new Geospatial Technology Competency Model: Bringing workforce needs into focus, *URISA Journal*, **22**, 55-72.

Droege, R.C., 1983, United States employment-service aptitude and interest testing for occupational exploration, *Journal of Employment Counseling*, **20**, 179-185.

Dyché, J., and E. Levy, 2006, *Customer Data Integration: Reaching a Single Version of the Truth*, Wiley, New York, 336 pp.

Dykes, J.A., A.M. MacEachren, and M.-J. Kraak, eds., 2005, *Exploring Geovisualization*, Elsevier, Amsterdam, 710 pp.

Estes, J.E., and J.R. Jensen, 1998, Development of remote sensing digital image processing systems and raster GIS, in *History of Geographic Information Systems*, T. Foresman, ed., Longman, New York, pp. 163-180.

Estrin, D., 2010, Participatory sensing: Applications and architecture, in *Proceedings of the 8th International Conference on Mobile Systems, Applications and Services (MobiSys'10)*, San Francisco, CA, June 15-18, Association for Computing Machinery, New York, pp. 3-4.

Fini, A., 2009, The technological dimension of a Massive Open Online Course: The case of the CCK08 course tools, *International Review of Research in Open and Distance Learning*, **10** (Special Issue).

Fischer, M., and A. Getis, 1997, *Recent Developments in Spatial Analysis*, Springer-Verlag, Berlin, 433 pp.

Fischer, M., and A. Getis, 2010, *Handbook of Applied Spatial Analysis: Software Tools, Methods and Applications*, Springer-Verlag, Heidelberg, 811 pp.

Flanagin, A.J. and M.J. Metzger, 2008, The credibility of volunteered geographic information, *GeoJournal*, **72**, 137-148.

Floberghagen, R., M. Fehringer, D. Lamarre, D. Muzi, B. Frommknecht, C. Steiger, J. Piñeiro, and A. da Costa, 2011, Mission design, operation and exploitation of the Gravity Field and Steady-state Ocean Circulation Explorer Mission, *Journal of Geodesy*, **85**, 749-758.

Foresman, T.W., T. Cary, T. Shupin, R. Eastman, J.E. Estes, K.K. Kemp, N. Faust, J.R. Jensen, and K. McGuire, 1997, The Remote Sensing Core Curriculum Program: An Internet resource for international education, *International Journal of Remote Sensing*, **52**, 294-300.

Fotheringham, A.S., and P. Rogerson, 1994, *Spatial Analysis and GIS*, Taylor & Francis, London, 281 pp.

Fotheringham, A.S., and P. Rogerson, 2009, *The SAGE Handbook of Spatial Analysis*, Sage, London, UK, 511 pp.

Freeman, L.C., 1977, A set of measures of centrality based on betweenness, *Sociometry*, **40**, 35-41.

Freeman, L.C., 2006, *The Development of Social Network Analysis*, Empirical Press, Vancouver, 205 pp.

Frost, T.S., J.M. Birkinshaw, and P.C. Ensign, 2002, Centers of excellence in multinational corporations, *Strategic Management Journal*, **23**, 997-1018.

Gasparikova, J., 2007, Is new economic research more qualitative? (Treatment on forecasting methods), *Journal of Economics*, **55**, 287-296.

Geiss, R., 2006, Asymmetric conflict structures, *International Review of the Red Cross*, **88**, 757-777.

Getkate, M., P. Hausdorf, and S.F. Cronshaw, 1992, Transnational validity generalization of employment tests from the United States to Canada, *Canadian Journal of Administrative Sciences*, **9**, 4324-4335.

Gewin, V., 2004, Mapping opportunities, *Nature*, **427**, 376-377.

Golde, C.M., and T.M. Dore, 2001, *At Cross Purposes: What the Experiences of Doctoral Students Reveal About Doctoral Education*, A report prepared for The Pew Charitable Trusts Philadelphia, PA, available at <www.phd-survey.org>.

Goldstone, J.A., R.H. Bates, D.L. Epstein, T.R. Gurr, M.B. Lustik, M.G. Marshall, J. Ulfelder, and M. Woodward, 2010, A global model for forecasting political instability, *American Journal of Political Science*, **54**, 190-208.

Goodchild, M., 1987, A spatial analytical perspective on Geographical Information Systems, *International Journal of Geographical Information Systems*, **1**, 31-45.

Goodchild, M.F., 1992, Geographical information science, *International Journal of Geographical Information Science*, **6**, 31-45.

Goodchild, M.F., 2006, Geographical information science fifteen years later, in *Classics from IJGIS: Twenty Years of the International Journal of Geographical Science and Systems*, P.F. Fisher, ed., CRC Press, Boca Raton, FL, pp. 199-204.

Goodchild, M.F., 2007, Citizens as sensors: The world of volunteered geography, *GeoJournal*, **69**, 211-221.

Goodchild, M., R. Haining, S. Wise, and 12 others, 1992, Integrating GIS and spatial data analysis—Problems and possibilities, *International Journal of Geographical Information Systems*, **6**, 407-423.

Hägerstrand, T., 1967, *Innovation Diffusion as a Spatial Process*, A. Pred, trans., University of Chicago Press, Chicago, 334 pp.

Haggett, P., and R.J. Chorley, 1969, *Network Analysis in Geography*, Edward Arnold, London, 348 pp.

Halevy, A., A. Rajaraman, and J. Ordille, 2006, Data integration: The teenage years, in *Proceedings of the 32nd International Conference on Very Large Data Bases*, Association for Computing Machinery, New York, pp. 9-16.

Hall, D.L., and J. Llinas, 1997, An introduction to multi-sensor data fusion, *Proceedings of the IEEE*, **85**, 6-23.

Hoh, B., T. Iwuchukwu, Q. Jacobson, M. Gruteser, A. Bayen, J-.C. Herrera, R. Herring, D. Work, M. Annavaram, and J. Ban, 2012, Enhancing privacy and accuracy in probe vehicle based traffic monitoring via virtual trip lines, *IEEE Transactions on Mobile Computing*, **11**, 849-864.

Hornsby, K., and M. Yuan, 2008, *Understanding Dynamics of Geographic Domains*, CRC Press, Boca Raton, FL, 240 pp.

Hosaka, T., 2008, Facebook asks users to translate for free Crowdsourcing aids company's aggressive worldwide expansion, MSNBC, available at <http://www.msnbc.msn.com/id/24205912/ns/technology_and_science-internet/t/facebook-asks-users-translate-free>.

Howarth, J.T., and D. Sinton, 2011, Sequencing spatial concepts in problem-based GIS instruction, *Procedia—Social and Behavioral Sciences*, **21**, 253-259.

Howe, J., 2006, The rise of crowdsourcing, *Wired Magazine*, June, available at <http://www.wired.com/wired/archive/14.06/crowds.html>.

Hyder, A., E. Shahbazian, and E. Waltz, eds., 2002, *Multisensor Fusion*, NATO Science Series, II. Mathematics, Physics and Chemistry, Vol. 70, Kluwer, Dordrecht, The Netherlands, 944 pp.

Jantsch, E., 1972, Towards interdisciplinarity and transdisciplinarity in education and innovation, in *Interdisciplinarity: Problems of Teaching and Research in Universities*, OECD/CERI, Paris, pp. 97-121.

Jensen, J.R., 2005, *Introductory Digital Image Processing: A Remote Sensing Perspective*, 3rd ed., Prentice Hall, Upper Saddle River, NJ, 526 pp.

Jensen, J.R., 2007, *Remote Sensing of the Environment: An Earth Resource Perspective*, 2nd ed., Prentice Hall, Upper Saddle River, NJ, 592 pp.

Jensen, J.R. and R.R. Jensen, 2012, *Introductory Geographic Information Systems*, Pearson/Prentice Hall, Upper Saddle River, NJ, 400 pp.

Jensen, J.R., J. Im, P. Hardin, and R.R. Jensen, 2009, Image classification, in *The SAGE Handbook of Remote Sensing*, T.A. Warner, M.D. Nellis, and G. Foody, eds., SAGE, London, pp. 269-281.

Kang, H., 2009, *Map Conflation: Analytical Conflation of Spatial Data from Municipal and Federal Government Agencies*, VDM Verlag, Saarbrücken, Germany, 252 pp.

Kates, R.W., 1987, The human environment: The road not taken, the road still beckoning, *Annals of the Association of American Geographers*, **77**, 525-534.

Keim, D., G. Andrienko, J.-D. Fekete, C. Görg, J. Kohlhammer, and G. Melançon, 2008, Visual analytics: Definition, process, and challenges, in *Information Visualization—Human-Centered Issues and Perspectives*, A. Kerren, J.T. Stasko, J.-D. Fekete, and C. North, eds., Lecture Notes in Computer Science, Vol. 4950, Springer, Berlin, pp. 154-175.

Kemp, K.K., and M.F. Goodchild, 1991, Developing a curriculum in GIS: The NCGIA core curriculum project, *Cartographica: The International Journal for Geographic Information and Geovisualization*, **28**, 39-54.

Kemp, K.K. and M.F. Goodchild, 1992, Evaluating a major innovation in higher education: The NCGIA core curriculum in GIS, *Journal of Geography in Higher Education*, **16**, 21-35.

Khachaturov, T., 1971, On methods and indices used in economic forecasting, *Problems of Economic Transition*, **14**, 3-24.

Kraus, K., 2004, *Photogrammetry: Geometry from Images and Laser Scans*, 2nd ed., de Gruyter, New York, 459 pp.

Krause, J., D.P. Croft, and R. James, 2007, Social network theory in the behavioural sciences: Potential applications, *Behavioral Ecology and Sociobiology*, **62**, 15-27.

Lélé, S., and R.B. Norgaard, 2004, Practicing interdisciplinarity, *BioScience*, **55**, 967-975.

Lenzerini, M., 2002, Data integration: A theoretical perspective, in *Proceedings of the ACM Symposium on Principles of Database Systems*, Association for Computing Machinery, New York, pp 233-246.

Lewin, T., 2012, Instruction for masses knocks down campus walls, *New York Times*, March 4.

Lillesand, T.M., R.W. Kiefer, and J.W. Chipman, 2008, *Remote Sensing and Image Interpretation*, John Wiley & Sons, New York, 756 pp.

Longley, P.A., M.F. Goodchild, D.J. Maguire, and D.W. Rhind, 2010, *Geographic Information Systems and Science*, 3rd ed., Wiley, New York, 560 pp.

Luo, Y., K. Ogle, C. Tucker, S. Fei, C. Gao, S. LaDeau, J.S. Clark, and D.S. Schimel, 2011, Ecological forecasting and data assimilation in a data-rich era, *Ecological Applications*, **21**, 1429-1442.

Lutgens, F.K., and E.J. Tarbuck, 1986, *The Atmosphere: An Introduction to Meteorology*, 3rd ed., Prentice Hall, Englewood Cliffs, NJ, 492 pp.

MacEachren, A.M., M. Gahegan, W. Pike, I. Brewer, G. Cai, E. Lengerich, and F. Hardisty, 2004, Geovisualization for knowledge construction and decision-support, *Computer Graphics and Applications*, **24**, 13-17.

MacLeod, M.N., 1919, Mapping from air photographs, *Geographical Journal*, **53**, 382-403.

Malys, S., J.A. Slater, R.W. Smith, L.E. Kunz, and S.C. Kenyon, 1997, Refinements to the World Geodetic System 1984, in *Proceedings of ION GPS-97*, the 10th International Technical Meeting of the Satellite Division of the Institute of Navigation, Kansas City, MO, September, Institute of Navigation, Washington, D.C., pp. 841-850.

Manyika, J., M. Chui, B. Brown, J. Bughin, R. Dobbs, C. Roxburgh, and A. Hung Byers, 2011, *Big Data: The Next Frontier for Innovation, Competition, and Productivity*, McKinsey Global Institute, 143 pp., available at <www.mckensey.com>.

Maus, S., 2010, An ellipsoidal harmonic representation of Earth's lithospheric magnetic field to degree and order 720, *Geochemistry Geophysics Geosystems*, **11**, Q06015, doi:10.1029/2010GC003026.

McCormick, B.H., T.A. DeFanti, and M.D. Brown, eds., 1987, Visualization in scientific computing, *Computer Graphics*, **21** (Special Issue).

McGlone, J.C., J.S. Brethel, and E.M. Mikhail, eds., 2004, *Manual of Photogrammetry*, 5th ed., American Society for Photogrammetry and Remote Sensing, Bethesda, MD, 1,168 pp.

McMaster, R., and S. McMaster, 2002, A history of twentieth century American academic cartography, *Cartography and Geographic Information Science*, **29**, 305-320.

Merrigan, M., E. Swift, R. Wong, and J. Saffel, 2002, A refinement to the World Geodetic System 1984 reference frame, in *Proceedings of ION GPS 2002*, 15th International Technical Meeting of the Satellite Division of the Institute of Navigation, Portland, OR, September, Institute of Navigation, Washington, D.C., pp. 1519-1529.

Mikhail, E.M., J.S Bethel, and J.C. McGlone, 2001, *Introduction to Modern Photogrammetry*, John Wiley & Sons, New York, 479 pp.

Mitchell, H.B., 2010a, *Image Fusion: Theories, Techniques and Applications*, Springer, Chennai, India, 200 pp.

Mitchell, H.B., 2010b, *Multi-Sensor Data Fusion: An Introduction*, Springer, Berlin, 282 pp.

Mondello, C., G.F. Hepner, and R.A. Williamson, 2004, 10-Year Industry Forecast, Phases I-III—Study Documentation, *Photogrammetric Engineering & Remote Sensing*, **January**, 5-58.

Mondello, C., G. Hepner, and R. Medina, 2006, ASP&RS ten-year remote sensing industry forecast Phase IV, *Photogrammetric Engineering & Remote Sensing*, **September**, 986-1000.

Mondello, C., G. Hepner, and R. Medina, 2008, ASP&RS ten-year remote sensing industry forecast Phase V, *Photogrammetric Engineering & Remote Sensing*, **November**, 1297-1305.

Monmonier, M.S., 1985, *Technological Transition in Cartography*, University of Wisconsin Press, Madison, WI, 282 pp.

Monti, F., 2010, Combining judgment and models, *Journal of Money, Credit & Banking*, **42**, 1641-1662.

Morabito, M., D.Z. Pavlinic, A. Crisci, V. Capecchi, S. Orlandini, and I.B. Mekjavic, 2011, Determining optimal clothing ensembles based on weather forecasts, with particular reference to outdoor winter military activities, *International Journal of Biometeorology*, **55**, 481-490.

Münkler, H., 2003, The wars of the 21st century, *International Review of the Red Cross*, **85**, 7-22.

Nerad, M., and J. Cerny, 1999, *PhDs: 10 Years Later Study*, Center for Innovation and Research in Graduate Education, available at <http://depts.washington.edu/cirgeweb/c/research/phd-career-path-surveys/phds-ten-years-later/>.

NRC (National Research Council), 1978, *Geodesy: Trends and Prospects*, National Academy Press, Washington, D.C., 86 pp.

NRC, 1994, *Federal Support of Basic Research in Institutions of Higher Learning*, National Academy Press, Washington, D.C., 98 pp.

NRC, 1995, *Reshaping the Graduate Education of Scientists and Engineers*, National Academy Press, Washington, D.C., 220 pp.

NRC, 1997, *Modeling and Simulation—Linking Entertainment and Defense*, National Academy Press, Washington, D.C., 196 pp.

NRC, 2000, *Forecasting Demand and Supply of Doctoral Scientists and Engineers: Report of a Workshop on Methodology*, National Academy Press, Washington, D.C., 104 pp.

NRC, 2006, *Learning to Think Spatially: GIS as a Support System in the K-12 Curriculum*, The National Academies Press, Washington, D.C., 332 pp.

NRC, 2007, *Elevation Data for Floodplain Mapping*, The National Academies Press, Washington, D.C., 151 pp.

NRC, 2010a, *New Research Directions for the National Geospatial-Intelligence Agency: Workshop Report*, The National Academies Press, Washington, D.C., 60 pp.

NRC, 2010b, *Persistent Forecasting of Disruptive Technologies*, The National Academies Press, Washington, D.C., 136 pp.

NRC, 2010c, *Precise Geodetic Infrastructure: National Requirements for a Shared Resource*, The National Academies Press, Washington, D.C., 142 pp.

NWS (National Weather Service), 2008, *National Strategic Training and Education Plan Process and Annual Implementation Plan*, National Weather Service Manual 20-102, available at <http://www.nws.noaa.gov/directives/sym/pd02001002curr.pdf>.

Nyquist, J., 2000, *Re-Envisioning the Ph.D. to Meet the Needs of the 21st Century*, available at <http://www.grad.washington.edu/envision/practices/index.html>.

Nyquist, J., L. Manning, and D.H. Wulff, 1999, On the road to becoming a professor: The graduate student experience, *Change: The Magazine of Higher Learning*, **31**, 18-27.

Omenn, G.S., 2006, Grand challenges and great opportunities in science, technology and public policy, *Science*, **314**, 1696-1704.

Ondrejka, R.J., 1997, Corona's invisible ASPRS partners, in *CORONA Between the Sun and the Earth. The First NRO Reconnaissance Eye in Space*, R.A. McDonald, ed., American Society for Photogrammetry and Remote Sensing, Bethesda, MD, pp. 153-158.

Pallas, A.M., 2001, Preparing education doctoral students for epistomological diversity, *Educational Researcher*, **30**, 6-11.

Paris, C.M., W. Lee, and P. Seery, 2010, The role of social media in promoting special events: Acceptance of Facebook 'events,' in *Information and Communication Technologies in Tourism 2010*, Springer, Vienna, pp. 531-541.

Perry, C., J. McIntyre, B. Starr, H. Schuster, and R. Habib, 2006, Cell phone tracking helped find al-Zarqawi U.S. military: Terrorist alive briefly after airstrike, *CNN*, June 10, available at <http://www.cnn.com/2006/WORLD/meast/06/09/iraq.al.zarqawi/>.

Pohl, C., and J.L. Van Genderen, 1998, Multi-sensor image fusion in remote sensing: Concepts, methods and applications, *International Journal of Remote Sensing*, **19**, 823-854.

Procter, R., M. Rouncefield, M. Poschen, Y. Lin, and A. Voss, 2011, Agile Project Management: A case study of a virtual research environment development project, *Computer Supported Cooperative Work*, **20**, 197-225.

Renslow, M., ed., 2012, *ASPRS Airborne Topographic Lidar Manual*, American Society for Photogrammetry & Remote Sensing, Bethesda, MD, 485 pp.

Ribarsky, W., B. Fisher, and W.M. Pottenger, 2009, Science of analytical reasoning, *Information Visualization*, **8**, 254-263.

Richelson, J.T., 1999, *The U.S. Intelligence Community*, 4th ed., Westview, Boulder, CO, 544 pp.

Saalfeld, A., 1988, Conflation: Automated map compilation, *International Journal of Geographical Information Science*, **2**, 217-228.

Schiermeier, Q., 2010, Weighing the world, *Nature*, **467**, 648-649.

Scholtz, J., K.A. Cook, M.A. Whiting, D. Lemon, and H. Greenblatt, 2009, Visual analytics technology transition progress, *Information Visualization*, **8**, 294-304.

Severing, A.J., 2011, Dealing with data: Training new scientists, Letter, *Science*, **331**, 1516.

Shekhar, S., and H. Xiong, eds., 2008, *Encyclopedia of GIS*, Springer, New York, 1377 pp.

Shekhar, S., M. Evans, J. Kang, and P. Mohan, 2011, Identifying patterns in spatial information: A survey of methods, *Wiley Interdisciplinary Reviews: Data Mining and Knowledge Discovery*, **1**, 193-214.

Sheth, A., and J. Larson, 1990, Federated database systems for managing distributed, heterogeneous, and autonomous databases, *ACM Computing Surveys*, **22**, 183-236.

Slocum, T.A., R.B. McMaster, F.C. Kessler, and H.H. Howard, 2009, *Thematic Cartography and Geovisualization*, Pearson/Prentice Hall, Upper Saddle River, NJ, 561 pp.

Snavley, N., S.M. Seitz, and R. Szeliski, 2006, Photo tourism: Exploring photo collections in 3D, *ACM Transactions on Graphics*, **25**, 835-846.

Solem, M., I. Cheung, and M.B. Schlemper, 2008, Skills in professional geography: An assessment of workforce needs and expectations, *The Professional Geographer*, **60**, 356-373.

Stapleton, G., G. Vitiello, and M. Sebillo, 2011, Guest editors' introduction, Special issue on visual analytics and visual semantics, *Journal of Visual Languages and Computing*, **22**, 171-172.

Stevens, M.J., and M.A. Campion, 1999, Staffing work teams: Development and validation of a selection test for teamwork settings, *Journal of Management*, **25**, 207-228.

Strober, M.H., 2006, Habits of the mind: Challenges for multidisciplinary engagement, *Social Epistemology*, **20**, 315-331.

Sweeney, W.C., 1924, *Military Intelligence: A New Weapon in War*, Frederick A. Stokes Co., New York, 259 pp.

Swift, E.R., 1994, Improved WGS 84 coordinates for the DMA and Air Force tracking sites, in *Proceedings of ION GPS-94*, 7th International Technical Meeting of the Satellite Division of the Institute of Navigation, Salt Lake City, UT, September 1994, Institute of Navigation, Washington, D.C., pp. 285-292.

Thomas, J.J., and K.A. Cook, eds., 2005, *Illuminating the Path. The Research and Development Agenda for Visual Analytics*, IEEE Computer Society, Los Alamitos, CA, 184 pp.

Thomas, J.J., and K.A. Cook, 2006, A visual analytics agenda, *IEEE Computer Graphics & Applications*, **26**, 10-13.

Torge, W. and J. Müller, 2012, *Geodesy*, 4th ed., Walter de Gruyter, Berlin, 433 pp.

Ture, S.A., 2004, Planning the future of the World Geodetic System 1984, in *Proceedings of IEEE Position Location and Navigation Symposium (PLANS) 2004*, Monterey, CA, April 26-29, IEEE, New York, pp. 639-648.

Urban, D., and T. Keitt, 2001, Landscape connectivity: A graph-theoretic perspective, *Ecology*, **82**, 1205-1218.

van Wijk, J.J., 2011, Guest editor's introduction: Special section on the IEEE Symposium on Visual Analytics Science and Technology (VAST), *IEEE Transactions on Visualization and Computer Graphics*, **17**, 5.

Vaníček, P., and E. Krakiwsky, 1986, *Geodesy: The Concepts*, 2nd ed., Elsevier, Amsterdam, 697 pp.

Wang, S., 2010, A cyberGIS framework for the synthesis of cyber-infrastructure, GIS and spatial analysis, *Annals of the Association of American Geographers*, **100**, 535-57.

Wang, X., J.X. Chen, D.B. Carr, B.S. Bell, and L.W. Pickle, 2002, Geographic statistics visualization: Web-based linked micromap plots, *Computing Science and Engineering*, **4**, 90-94.

Wasserman, S., and K. Faust, 1994, *Social Network Analysis: Methods and Applications*, Cambridge University Press, Cambridge, 825 pp.

Weinberger, S., 2011, Web of war, *Nature*, **471**, 566-568.

White, F.E., 1999, Managing data fusion systems in joint and coalition warfare, in *Proceedings of the 33rd Asilomar Conference on Signals, Systems, and Computers*, IEEE, New York, pp. 412-415.

Wiley, B., D. Craig, D. Manning, J. Novak, R. Taylor, and L. Weingarth, 2006, NGA's role in GPS, in *Proceedings of ION GNSS 2006*, 19th International Technical Meeting of the Satellite Division of the Institute of Navigation, Fort Worth, TX, September 2006, Institute of Navigation, Washington, D.C., pp. 2111-2119.

Wilkins, A.L., and W.G. Ouchi, 1983, Efficient cultures: Exploring the relationship between culture and organizational performance, *Administrative Science Quarterly*, **28**, 468-481.

Wilson, J.W., Y. Feng, M. Chen, and R.D. Roberts, 2010, Nowcasting challenges during the Beijing Olympics: Successes, failures, and implications for future nowcasting systems, *Weather & Forecasting*, **25**, 1691-1714.

Winchester, S., 1998, *The Professor and the Madman: A Tale of Murder, Insanity, and the Making of the Oxford English Dictionary*, Harper Collins, New York, 242 pp.

Yiannakis, A., M.J.P. Selby, J. Douvis, and J.Y. Han, 2006, Forecasting in sport: The power of social context—A time series analysis with English Premier League Soccer, *International Review for the Sociology of Sport*, **41**, 89-115.

Zook, M., M. Graham, T. Shelton, and S. Gorman, 2010, Volunteered geographic information and crowdsourcing disaster relief: A case study of the Haitian earthquake, *World Medical & Health Policy*, **2**(2), Article 2.

Zyda, M., 2009, Computer science in the conceptual age, *Communications of the ACM*, **52**, 66-72.

Appendix A

Example University Programs and Curricula

An overview of the academic courses and skills needed for the 10 core and emerging areas is given in Chapters 2 and 3, and example academic programs in these areas are described in Chapter 6. This appendix provides supporting information, including example university programs associated with Chapters 2 and 3 (Tables A.1–A.10) and example university curricula for degree and certificate programs discussed in Chapter 6 (Tables A.11–A.15). The committee selected the example programs and curricula based on the following criteria:

- longevity of the program,
- critical mass of instructors,

- number of students graduated in the academic and skill areas sought by NGA,
- caliber of instructors, and
- a curricular focus that allows the types of problem solving and analysis needed by NGA.

For the university programs in the core areas, the committee used its expert judgment to choose 10 to 15 examples from a longer list of qualified programs. For the university programs in emerging areas, where university degree programs do not exist, the committee chose 5 to 10 programs that offer reasonably comprehensive coursework and relevant skills.

UNIVERSITY PROGRAMS IN CORE AND EMERGING AREAS

Example Programs in Core Areas

TABLE A.1 Example University Programs in Geodesy and Geophysics

University	Department	Concentration/Track	Degree
Geodesy			
California State University, Fresno	Civil and Geomatics Engineering	Geomatics engineering	B.S.
Ferris State University	College of Engineering Technology	Surveying engineering	B.S.
Florida Atlantic University	Civil, Environmental and Geomatics Engineering	Geomatics engineering	B.S.
Massachusetts Institute of Technology	Earth, Atmospheric and Planetary Sciences	Geodesy	M.S., Ph.D.
Ohio State University	Civil, Environmental, and Geodetic Engineering; and Division of Geodetic Science	Geospatial and geodetic engineering; geodetic science	M.S., Ph.D.
Oregon Institute of Technology	Geomatics	Surveying	B.S.
Pennsylvania State University	College of Engineering	Surveying engineering	B.S.
Texas A&M University, Corpus Christi	School of Engineering and Computing Sciences	Geomatics	B.S., M.S.
University of Alaska, Anchorage	Geomatics	Geomatics	B.S.
University of Colorado, Boulder	Aerospace Engineering Sciences	Geodesy	M.S., Ph.D.
University of Florida	School of Forest Resources and Conservation	Geomatics	B.S., M.S., Ph.D.
University of Maine	School of Engineering Technology	Surveying Engineering Technology	B.S.
University of Texas, Austin	Center for Space Research	Geodesy	M.S., Ph.D.
Geophysics			
California Institute of Technology	Division of Geological and Planetary Sciences	Geophysics	B.S., M.S., Ph.D.
Columbia University	Earth and Environmental Sciences	Geoscience	B.S., Ph.D.
Harvard University	Earth and Planetary Sciences	Geophysics	B.A., Ph.D.
Massachusetts Institute of Technology	Earth, Atmospheric and Planetary Sciences	Geoscience	B.S., M.S., Ph.D., D.Sc.
Princeton University	Geosciences	Geophysics	B.A., Ph.D.
Stanford University	Geophysics	Geophysics	B.S., M.S., Ph.D.
University of California, Berkeley	Earth and Planetary Science	Geophysics	B.S., M.A., Ph.D.
University of California, Santa Cruz	Earth and Planetary Sciences	Geophysics	B.S., M.S., Ph.D.
University of Southern California	Earth Sciences	Geophysics	B.S., M.S., Ph.D.
University of Texas, Austin	Institute for Geophysics	Geophysics	M.S., Ph.D.
University of Washington	Earth and Space Sciences	Geophysics	B.S., M.S., Ph.D.

Approximate number of schools: 20 for geodesy and 60 for geophysics.

TABLE A.2 University Programs Offering Some Photogrammetry

University	Department	Concentration/Track	Degree
Ohio State University	Geodetic Science	Photogrammetry	M.S., Ph.D.
Purdue University	Geomatics Engineering	Photogrammetry	M.S., Ph.D.
University of Florida	Geomatics	Geomatics, Photogrammetry	B.S., M.S., Ph.D.[a]
Ferris State University	School of CEEMS	Surveying Engineering	B.S.[b]
California State University, Fresno	Civil and Geomatics Engineering	Geomatics Engineering, Photogrammetry	B.S., M.S.[a]
California State Polytechnic University, Pomona	Civil Engineering	Geospatial Engineering	B.S.[b]
New Mexico State University	Engineering Technology and Surveying Engineering	Surveying Engineering	B.S.[b]
Oregon Institute of Technology	Geomatics	Geomatics	B.S.[b]
Texas A & M University, Corpus Christie	Geographic Information Science and Geospatial Surveying Engineering	Geomatics, Geospatial Surveying	B.S., M.S.[b]
Pennsylvania State University, Wilkes-Barre	Surveying Engineering	Surveying Engineering	B.S.[b]
University of Alaska, Anchorage	Geomatics	Geomatics	B.S.[b]
George Mason University	Geography and Geoinformation Science	Geography, Geoinformation Science	M.S., Ph.D.[c]
East Tennessee State University	Surveying and Mapping	GIS and Photogrammetry	B.S.[b]

Approximate number of schools: 15.

[a] B.S. includes some courses in photogrammetry; graduate degree has a concentration in photogrammetry.

[b] Includes some courses in photogrammetry.

[c] One introductory course in photogrammetry taught by a non-faculty member (from industry).

TABLE A.3 Example Universities with a Remote Sensing-Related Track or Degree

University	Department	Concentration/Track	Degree
Air Force Institute of Technology	Engineering Physics	Engineering	M.S., Ph.D.
Boston University	Geography	Remote sensing	B.A., M.A., Ph.D.
Clark University	Geography	Remote sensing	B.A., M.A., Ph.D.
George Mason University	Geography and GeoInformation Science	Remote sensing	B.A., MS, Ph.D.
Naval Post Graduate School	Information Science	Engineering	M.S., Ph.D.
Pennsylvania State University	Geography	Remote sensing	B.S., M.S., Ph.D.
Rochester Institute of Technology	Center for Imaging Science	Remote sensing	B.S., M.S., Ph.D.
University of California, Santa Barbara	Geography	Remote sensing	B.A., M.A., Ph.D.
University of Colorado, Boulder	Geography	Remote sensing	B.A., M.A., Ph.D.
University of Maryland	Geography	Remote sensing	B.S., M.A., Ph.D.
University of Michigan	Geoscience and Remote Sensing	Engineering	B.S., M.S., Ph.D.
University of South Carolina	Geography	Remote sensing	B.A., M.S., Ph.D.
University of Utah	Geography	Remote sensing	B.S., M.S., Ph.D.
University of New Hampshire	Forestry	Remote sensing	B.A., M.S., Ph.D.
University of Montana	Ecosystem and Conservation Science	Remote sensing	B.S., M.S., Ph.D.
University of Georgia	Geography	Remote sensing	B.A., M.A., Ph.D.

Approximate number of schools: 63.

TABLE A.4 Example Universities with a Cartography Track or Degree

University	Department	Concentration/Track	Degree
University of Arkansas	Geosciences	Cartography/remote sensing	B.A., M.A.
University of Colorado, Boulder	Geography	GIS and cartography	B.A., M.A., Ph.D.
Southern Illinois University, Edwardsville	Geography	GIS and cartography	B.A., M.S.
Salem State College	Geography	Cartography and GIS	B.S., M.S.
Frostburg State University	Geography	Mapping sciences	B.S.
University of Nebraska, Lincoln	Geography	GIS, cartography, and remote sensing	M.S.
State University of New York, Binghamton	Geography	Cartography and GIS	M.A.
Kent State University	Geography	GIS and cartography	B.A. minor
East Central University	Cartography and Geography	Geotechniques	B.S.
Indiana University of Pennsylvania	Geography and Regional Planning	GIS cartographer	B.A., M.S.
George Mason University	Geography and GeoInformation Science	Geographic and cartographic sciences	M.S.
University of Washington	Geography	GIS mapping and society	B.A.
University of Wisconsin, Madison	Geography	Cartography and geographic information science	B.S., M.S.
University of Wisconsin, River Falls	Geography and Mapping Sciences	GIS and cartography	B.A. minor

Approximate number of schools: 35.

TABLE A.5 Example Universities with Degree Tracks in GIS and Geographic Information Science

University	Department	Concentration/Track	Degree
Arizona State University	School of Geographical Sciences and Urban Planning	GIS-Spatial Analysis	M.A.
Clark University	International, Development, Community and Environment; and School of Geography	Geographic Information Science for Development and Environment	M.A.
Pennsylvania State University	Geography; and John A. Dutton e-Education Institute	GIS	M.
University of California, Santa Barbara	Geography	Modeling, measurement, and computation	M.A., M.S.
University of Colorado, Boulder	Geography	GIS	M.A.
University of Minnesota	Geography	GIS	M.
University of Pennsylvania	School of Design	Urban spatial analytics	M.
University of Redlands	GIS	GIS	M.S.
University of South Carolina	Geography	Geographic information science	M.A., M.S.
University of Southern California	Dana and David Dornsife College of Letters, Arts and Sciences	Geographic information science and technology	M.
University of Washington	Professional and Continuing Education	GIS	Prof. M.
State University of New York, Buffalo	Geography	Geographic information systems and science	M.A., M.S.

Approximate number of schools: 189 degree programs in GIS and more than 400 community colleges and technical schools that offer some form of training in geospatial technologies. See <http://www.urisa.org/career/colleges>.

Example Programs in Emerging Areas

TABLE A.6 Example Universities with Courses Relevant to GEOINT Fusion

University	Department or Research Unit	Concentration/Track	Academic Structure
University of Southern California	Computer Science	Database interoperability	Graduate
Georgia Institute of Technology	Electrical Engineering	Multisensor data fusion	Graduate
University of California, Santa Barbara	Geography	Geographic information science: Map conflation	Undergraduate
Purdue University	Civil Engineering	Geomatics: Image fusion	Undergraduate
Johns Hopkins University	Computer Science	Semantic web	Graduate
Pennsylvania State University	Information Science and Technology	Multisensor data fusion	Graduate
State University of New York, Buffalo	Industrial and Systems Engineering	Multisource information fusion	Graduate

Approximate number of schools: 12 offering courses covering some aspects of data fusion.

TABLE A.7 Example Universities with Courses Relevant to Crowdsourcing

University	Department or Research Unit	Concentration/Track	Academic Structure
University of California, Los Angeles	Center for Embedded Networks Systems	Participatory sensing	Graduate
University of California, Berkeley	Algorithms, Machines, and People Lab	Large-scale data analytics	Graduate
Massachusetts Institute of Technology	Computer Science and Artificial Intelligence Laboratory	Systems for crowdsourcing	Graduate
Rutgers University	Wireless Information Networks Laboratory	Privacy, security	Graduate
University of Pennsylvania	General Robotics, Automation, Sensing and Perception Laboratory	Sensor networks	Graduate

Approximate number of schools: fewer than 10.

TABLE A.8 Example Universities with Courses Relevant to Human Geography

University	Department or Research Unit	Concentration/Track	Academic Structure
Carnegie Mellon University	Institute for Software Research	Computation, organizations, and society	Graduate
University of California, Irvine	Sociology	Social networks	Undergraduate, graduate
Duke University	Sociology	Social networks	Undergraduate, graduate
University of Arizona	Sociology	Social networks	Undergraduate, graduate

Approximate number of schools: fewer than 10.

TABLE A.9 Example Universities with Courses Relevant to Visual Analytics

University	Department or Research Unit	Concentration/Track	Academic Structure
Purdue University	Visual Analytics for Command, Control, and Interoperability Environments	Visualization, data sciences	Graduate
University of North Carolina, Charlotte	Charlotte Visualization Center	Graphics, visualization	Undergraduate, graduate
Georgia Institute of Technology	Graphics, Visualization, and Usability Center	Computer graphics, visualization, human-computer interface	Undergraduate, graduate
University of Washington	Pacific Rim Visualization and Analytics Center	Visual analytic systems	Undergraduate, graduate
University of Massachusetts, Lowell	Institute for Visualization and Perception Research	Visualization technologies, visual analytics	Undergraduate, graduate
Stanford University	Stanford Visualization Group	Data analysis, visualization	Graduate
University of California, Santa Barbara	Media Arts and Technology	Media arts and technology	Graduate

Approximate number of schools: 15.

TABLE A.10 Example Universities with Courses Relevant to Forecasting

University	Department or Research Unit	Concentration/Track	Academic Structure
University of Oklahoma	School of Meteorology	Weather forecasting	Undergraduate, graduate
University of Washington	Atmospheric Sciences	Weather forecasting	Undergraduate, graduate
Arizona State University	School of Geographical Sciences and Urban Planning	Spatial statistics, spatial modeling, and econometrics	Undergraduate, graduate
Ohio State University	Statistics	Spatial statistics, spatial modeling, and econometrics	Undergraduate, graduate
University of Texas, Dallas	School of Economic, Political, and Policy Sciences	Spatial statistics, spatial modeling, and econometrics	Undergraduate, graduate
Pennsylvania State University	Geography	Spatial statistics, spatial modeling, and econometrics	Undergraduate, graduate
Carnegie Mellon University	Institute for Software Research	Agent-based modeling	Undergraduate, graduate
George Mason University	Computer Science	Agent-based modeling	Undergraduate, graduate
University of Michigan	Center for the Study of Complex Systems	Agent-based modeling	Undergraduate, graduate
Massachusetts Institute of Technology	Mechanical Engineering	Complex systems and economics	Undergraduate, graduate
Northwestern University	Institute on Complex Systems	Complex systems and economics	Undergraduate, graduate
Harvard University	Economics	Economics	Undergraduate, graduate
Princeton University	Economics	Economics	Undergraduate, graduate
University of Chicago	Economics	Economics	Undergraduate, graduate

Approximate number of schools: more than 100. Most large universities offer courses in economics, computer science, mathematics, statistics, and social sciences that have analytical and modeling components relevant to prediction and forecasting.

EXAMPLE CURRICULA

TABLE A.11 Coursework for an Undergraduate Degree in Geographic Information Science at the University of Colorado

Course[a]	Title	Description
Two or three of the following (8–12 credits)		
GEOG 2053 (4 credits)	Mapping a changing world	Overviews the vital role cartography plays in modern society and contemporary science. Fundamentals of reading and creating maps for research and enjoyment
GEOG 3023 (4 credits)	Statistics for earth sciences	Introduces parametric and distribution-free statistics, emphasizes applications to earth science problems. Not open to students who have taken a college-level statistics course. Restricted to junior and senior geography, geology, and environmental studies majors
GEOG 3053 (4 credits)	Cartography 1: Visualization and information design	Fundamentals of cartography—the science and art of map design. Restricted to junior and senior geography and environmental studies majors. Recommended GEOG 3023 (may be taken concurrently)
All of the following (9 credits)		
GEOG 4023 (3 credits)[b]	Introduction to quantitative methods in human geography	Introduces fundamental statistical and quantitative modeling techniques widely used in geography today. Emphasizes geographic examples and spatial problems, using statistical routines now available on most computers. Prereq GEOG 3023 or equivalent
GEOG 4033 (2 credits)[b]	Quantitative methods in geography laboratory	Introduces the use of personal computers and statistical software in geographical analysis. Corequisite GEOG 4023
GEOG 4103 (4 credits)	Introduction to geographic information science	Use of tools and databases specifically designed for spatial data. Covers data management and procedures for transforming, storing/retrieving, and analyzing geographic data. Restricted to junior and senior geography and environmental studies majors. Prerequisites GEOG 3023 and GEOG 3053
One to four of the following (4–16 credits)		
GEOG 4043 (4 credits)	Cartography 2: Interactive and multimedia mapping	Interactive, multimedia, animated, and Web-based cartography stressing the important role digital cartography plays in cyberspace. Focuses on principles of effective cartographic design in multimedia and hypertext environments. Prerequisite GEOG 3053
GEOG 4093 (4 credits)	Remote sensing of the environment	Acquisition and interpretation of environmental data by remote sensing. Topics include theory and sensors as well as manual and computerized interpretation methods. Stresses infrared and microwave portions of the spectrum
GEOG 4110 (4 credits)	Advanced remote sensing	Extends basic concepts and skills of image processing and physics of remote sensors, with deeper examination of image analysis techniques for extracting the maximum amount of information. Prerequisite GEOG 4093
GEOG 4203 (4 credits)	Geographic information science: Modeling applications	Develops GIS models for human and environmental applications, grid and vector data models, tesselated and hierarchical data structures, terrain representation, linear and areal interpolation and kriging. Students work in small group to design, implement, and run GIS models. Prerequisite GEOG 4103/5103, GEOG 3023 (or equivalent) or instructor consent
GEOG 4303 (4 credits)	Geographic information science: Programming	Introduces the use of Python programming to undertake advanced spatial analysis tasks within a GIS environment. Prerequisite GEOG 4103/5103, GEOG 3023 (or equivalent) or instructor consent
GEOG 4xxx (4 credits)	Spatial statistics (under development)	Involves the quantitative analysis of spatial data and statistical modeling of spatial variability and uncertainty. Topics may include point pattern analysis, model-based geostatistics, semivariogram analysis, validation methods and simulation
GEOG 5113 (4 credits)	Advanced spatial topics in GIS	Graduate seminar; topics vary

SOURCE: <http://geography.colorado.edu/undergrad_program/areas_of_concentration/geographic_information_science>.

[a] All 4-credit courses require 45 hours in lecture and 30–45 hours in laboratory.

[b] GEOG 4023 and 4033 must be taken concurrently and together require 45 hours in lecture and 45 hours in laboratory.

TABLE A.12 Coursework for a Master's Degree in Geographic and Cartographic Sciences at George Mason University

Course	Title	Description
Required courses (12 credits)[a]		
GGS 553 (3 credits)	Geographic Information System	Sources of digital geospatial data; and methods of input, storage, display, and processing of spatial data for geographic analysis using GIS. Lectures, hands-on exercises familiarize students with current technology
GGS 579 (3 credits)	Remote sensing	Examines use of various types and combinations of electromagnetic energy to obtain spatial information. Concentrates on nonphotographic and spaceborne remote sensing platforms and sensors. Examines essential operational parameters for existing and future systems and strategies for visual extraction of features
GGS 560 (3 credits)	Quantitative methods	Survey of quantitative methods commonly used in geographic research. Emphasizes spatial analysis techniques
GGS 689 (3 credits)	Seminar in geographic thought and methodology	Includes historical development of geographic thought and current philosophy of geography; rationale for various subfields; and geographic research techniques and methods of analysis
Electives (21-24 credits)[b]		
GGS 503 (3 credits)	Problems in environmental management	Case studies of effects of human activities on atmospheric, hydrologic, geomorphic, and biotic processes
GGS 505 (3 credits)	Transportation geography	Structure, principles, location, and development of world transportation. Critical role of transportation in moving people, goods, and ideas at international, national, regional, and urban levels
GGS 525 (3 credits)	Economics of human/ environmental interactions	Advanced topics in environmental, natural resource, and ecological economics for noneconomist. Emphasizes sustainability, intergenerational equity, and economic-ecological feedbacks
GGS 531 (3 credits)	Land-use modeling techniques and applications	Survey of literature on spatially explicit empirical models of land-use change. Hands-on experience developing and running simple models. Techniques covered include statistical models, mathematical programming models, cellular automata, agent-based models, and integrated models
GGS 533 (1-6 credits)	Issues in regional geography	Geographical study of particular region or relevant regional issue
GGS 540 (3 credits)	Medical geography	Spatial approaches to study of health and disease. Topics include disease ecology and diffusion, and geographic perspectives on improving health care delivery
GGS 551 (3 credits)	Thematic cartography	Analyzes nature of perceptual organization and visual systems in thematic map communication portrayal, graphic handling, and data analysis
GGS 554 (3 credits)	History of cartography	History of cartographic portrayal of Earth from ancient times through 19th century, emphasizing interrelation of human culture, technological development, and geographical knowledge as reflected in maps
GGS 562 (3 credits)	Photogrammetry	Treatment of photogrammetric problems, including least-squares adjustments, image coordination refinements, colinearity equation, resection, relative orientation, and analytic aerotriangulation
GGS 563 (3 credits)	Advanced Geographic Information Systems	Discusses advanced GIS concepts including spatial data structure, spatial analysis, programming data fusion, Internet components, and spatial database management. Hands-on activities demonstrate concepts and specific applications in both cultural and physical geography
GGS 579 (3 credits)	Digital remote sensing	Examines use of various types and combinations of electromagnetic energy to obtain spatial information. Concentrates on nonphotographic and spaceborne remote sensing platforms and sensors. Examines essential operational parameters for existing and future systems and strategies for visual extraction of features
GGS 581 (3 credits)	World food and population	Topics include maldistribution of population, regional disparities in growth rates and income distribution, food production, and world hunger. Discusses population policies, with emphasis on Third World countries
GGS 590 (1-3 credits)	Geography of insurgency	Special topics seminar which analyzes topics of immediate interest in political unrest and insurgency
GGS 590 (1-3 credits)	Political geography	Special topics seminar which analyzes topics of immediate interest in political policies and political behavior
GGS 590 (1-3 credits)	GIS for business	Special topics seminar which analyzes topics of immediate interest in business related applications of GIS such as enterprise GIS, GIS for real estate, location analysis and marketing
GGS 590 (1-3 credits)	GIS for the environment	Special topics seminar which analyzes topics of immediate interest in GIS and environmental modeling, conservation, and sustainability
GGS 605 (3 credits)	Socioeconomic applications of GIS	Provides those working with spatially referenced data the technical skills to use GIS to conduct spatial analyses on socioeconomic phenomena related to labor, retail, and real estate markets. Introduces and emphasizes the development of technical and methodological skills to understand the potential and the pitfalls of using GIS for spatial analyses of socioeconomic phenomena
GGS 631 (3 credits)	Spatial agent-based models of human environment interactions	Discusses key challenges in spatial modeling of human-environment interactions. Reviews agent-based modeling applications in urban and rural interactions, agriculture, forestry, and other areas. Hands-on development of simple agent-based models and investigation of linkages between GIS and agent-based models

continued

TABLE A.12 Continued

Course	Title	Description
GGS 650 (3 credits)	Introduction to GIS programming	Introduction to programming methods and their application to Geographic Information Systems, including the fundamentals of object-oriented programming and GIS-specific data structures and algorithms. Employs an object-oriented language such as Visual Basic.Net, and existing freeware and commercial GIS libraries. Topics covered include variables, arrays, control structures, objects and classes, raster and vector data structures, spatial algorithms, and spatial indexing methods
GGS 653 (3 credits)	Geographic information analysis	Explores existing and potential capabilities of geographic information systems in conducting spatial analysis and modeling
GGS 655 (3 credits)	Map design	Advanced examination of principles of map design, including discussions of map design research
GGS 656 (3 credits)	The hydrosphere	Covers components and transfer processes in hydrosphere, which consists of aqueous envelope of Earth including oceans, lakes, rivers, snow, ice, glaciers, soil moisture, groundwater, and atmospheric water vapor
GGS 658 (3 credits)	Terrain mapping	Covers fundamental methods of digitally representing terrain data, major technologies, and programs for generating terrain data; methods for quantifying terrain error and assessing terrain data quality; and a variety of applications
GGS 661 (3 credits)	Map projections and coordinate systems	Covers development of various map projections and coordinate systems, property analysis, distortions, and applications
GGS 664 (3 credits)	Spatial data structures	Studies spatial data structures and their application in digital cartography, geographic information systems, and image-processing systems. Examines raster and vector data structures, and attribution schemes and topological models. Includes data transformation, information loss, data quality, and the role of metadata

SOURCE: <http://cos.gmu.edu/academics/graduate/ms/geographic-and-cartographic-sciences-ms>.

[a] 3 semester credits translates to 45 hours in class.

[b] 18 credits of electives plus 3-6 credits GSS 799 Thesis Writing for the thesis option, and 24 credits of electives for the nonthesis option.

TABLE A.13 Coursework for an Online Master's Degree in GIS at Penn State World Campus

Course[a]	Title	Description
Required courses (23-26 credits)		
GEOG 482 (2 credits)	The nature of geographic information	Orientation to the properties of geographic data and the practice of distance learning
GEOG 483 (3 credits)	Problem solving with GIS	How geographic information systems facilitate data analysis and communication to address common geographic problems
GEOG 484 (3 credits)	GIS database development	Database design, creation, and maintenance, and data integration using desktop GIS software
GEOG 583 (3 credits)	Geospatial system analysis and design	Systematic approach to requirements acquisition, specification, design and implementation of geospatial information systems
GEOG 584 (3 credits)	Geospatial technology project management	Principles of effective project management applied to the design and implementation of geospatial information systems
GEOG 586 (3 credits)	Geographical information analysis	Choosing and applying analytical methods for geospatial data, including point pattern analysis, interpolation, surface analysis, overlay analysis, and spatial autocorrelation
GEOG 596A (3-9 credits)	Individual studies—Peer review	Preparation and presentation of a proposal for an individual capstone project
GEOG 596B (3-9 credits)	Individual studies—Capstone project	Preparation and delivery of a formal professional presentation of the results of an individual capstone project
Electives (minimum 9 credits)		
GEOG 485 (3 credits)	GIS programming and customization	Customizing GIS software to extend its built-in functionality and to automate repetitive tasks
GEOG 486 (3 credits)	Cartography and visualization	Theory and practice of cartographic design, emphasizing effective visual thinking and visual communication with geographic information systems
GEOG 487 (3 credits)	Environmental applications of GIS	Simulated internship experience in which students play the role of GIS analysts in an environmental consultancy
GEOG 488 (3 credits)	Acquiring and integrating geospatial data	Advanced technical, legal, ethical, and institutional problems related to data acquisition for geospatial information systems
GEOG 489 (3 credits)	GIS application development	Advanced topics in GIS customization, including the Systems Development Life Cycle, packaging and deployment, and consuming Web services
GEOG 497D (3 credits)	Lidar technology and applications	Understanding lidar systems; design, operation, data processing techniques, and product generation to address typical application scenarios faced by the geospatial professional
GEOG 587 (3 credits)	Conservation GIS	Conservation GIS applies geospatial problem solving to ecological research and resource management issues to enhance conservation planning
GEOG 588 (3 credits)	Planning GIS for emergency management	Requirements analysis and proposal writing to plan and implement GIS solutions supporting emergency management activities of government agencies and contractors
GEOG 597K (3 credits)	GIS for analysis of health	The role of geographic information systems in understanding disease, including relevant spatial analysis and cartographic visualization techniques
GEOG 861 (1 credit)	Map projections for geospatial professionals	Cultivates a working knowledge of map projections that professionals need to process geospatial data effectively for mapping and analysis
GEOG 862 or GEOG 497I (3 credits)	GPS and GNSS for geospatial professionals	Cultivates a working knowledge of current and future capabilities of GPS and the emerging Global Navigation Satellite System
GEOG 863 or GEOG 497J (3 credits)	GIS mashups for geospatial professionals	Cultivates a working knowledge of how and why geospatial professionals develop web mapping applications that combine data from multiple sources
GEOG 864 or GEOG 598E (2 credits)	Professionalism in GIS&T	Prepares current and aspiring professionals to recognize, analyze, and address ethical problems in the geographic information science and technology field
GEOG 897G (3 credits)	Trends in geospatial technology	Developing lifelong learning skills to take advantage of the changing tools of geospatial technology
STAT 480 (1 credit)	Introduction to SAS	Introduction to SAS with emphasis on reading, manipulating, and summarizing data
STAT 505[b] (3 credits)	Applied multivariate statistical analysis	Analysis of multivariate data; T-squared tests; partial correlation; discrimination; MANOVA; cluster analysis; regression; growth curves; factor analysis; principal components; canonical correlations

SOURCE: <http://www.worldcampus.psu.edu/degrees-and-certificates/geographic-information-systems-gis-masters/course-list>.
[a] The M.S. degree requires 35 credits and is expected to take 3 years full time to complete.
[b] Elective for a master's degree in GIS only.

CERTIFICATES

TABLE A.14 Selected Institutions That Offer Cartography, GIS, and Remote Sensing-Related Certificates

Institution, Department	Certificate Title	Course Requirements
California State University, East Bay, Geography	Certificate in Cartography and GIS	20 hours
George Mason University, Geography and GeoInformation Science	Graduate Certificate in Remote Sensing	15 hours
Georgia Tech, School of Earth and Atmosphere	Certificate Program in Remote Sensing	12 hours
Humbolt State, Forestry and Wildland Resources	Certificate in GIS and Remote Sensing	5 courses
Institute of Geoinformatics and Remote Sensing	Post Graduate Certificate in GIS and Remote Sensing	6 months; 150 hours
University of Twente, International Institute for Geo-Information Science and Earth Observation	Hyperspectral Remote Sensing Certificate	6 weeks
	Remote Sensing and GIS Geology Exploration Certificate	9 weeks
	Principles of Remote Sensing Certificate	9 weeks
Oregon State University, College of Forestry	GIScience Certificate with Emphasis in Remote Sensing	19 hours
Symbosis Institute of Geoinformatics	Certificate Course in Photogrammetry and Remote Sensing	2 months
Mississippi State University, Division of Academic Outreach	Geospatial and Remote Sensing Technology Certificate	15 hours
Naval Postgraduate School	Space Systems Certificate	4 courses
Northeastern University, Professional Studies	Graduate Certificate in Remote Sensing	6 courses
Rutgers University, Geography	Geospatial Information Science Certificate	12 hours
San Jose State University, Geography	Certificate in GIScience with Specialization in Remote Sensing	18 hours
West Virginia University, Geography	Certificate in GIS and Remote Sensing	15 hours
University of Colorado, Boulder, Aerospace Engineering	Remote Sensing Certificate	4 courses
University of California, Davis, Center for Spatial Technologies and Remote Sensing	Base, Intermediate, and Advanced Certificates in Remote Sensing	not available
University of Maryland, Professional Studies	Graduate Certificate in GIS	12 hours
University of Michigan, Dearborn, Natural Sciences	Certificate in GIS and Remote Sensing	16 hours
University of Mississippi, School of Law	Certificate in Remote Sensing, Air, and Space Law	27 hours
University of New Orleans, Geography	Remote Sensing and GIS Certification	4 courses
University of Omaha, Geography	Graduate GIScience Certificate	17 hours
University of Texas, Dallas, Economic Policy	Graduate Certificate in Remote Sensing	15 hours
	Graduate Certificate in Geospatial Intelligence	15 hours
University of Texas, Arlington, Geology	Certification in Remote Sensing, GPS and GIS	15 hours
University of Utah, Geography	Certificate in GIS with Emphasis in Remote Sensing	22 hours
St. Louis University, Environmental Sciences	Graduate Certificate in Advances in Remote Sensing and GIS	15 hours
Texas A&M University, Geography	Remote Sensing Certification	4 courses; 12 hours
York University	GIS and Remote Sensing Certificate	not available
Webster University, Business and Technology	Graduate Certificate in Remote Sensing Analysis and GIS	18 hours

SOURCES: Association of American Geographers, Environmental Systems Research Institute, and the Urban and Regional Science Association.

TABLE A.15 Course Requirements for a Certificate in GIS with an Emphasis in Remote Sensing at the Department of Geography, University of Utah

Course Number	Title	Hours in Class[a]
GEOG 3020	Geographical analysis	45
GEOG 3110	The Earth from space: Remote sensing of the environment	45
GEOG 3140	Introduction to Geographic Information Systems	45
GEOG 5110	Environmental analysis through remote sensing	45
GEOG 5120	Environmental optics	45
GEOG 5130	Advanced remote sensing applications	45
One of the following:		
CS 1000	Engineering computing	45
CS 1020	Introduction to programming in C++	45
CS 1021	Introduction to programming in Java	45
CS 1410	Introduction to computer science I	60
CS 2000	Introduction to program design in C	60
Total hours in class		315 or 330

SOURCE: <http://www.geog.utah.edu/giscert/rs_track.html>.
[a] 3 semester credit hours = 45 hours in class.

Appendix B

Job Descriptions of NGA Scientists and Analysts

Below are descriptions of NGA's science and geospatial intelligence (GEOINT) positions and educational requirements (Table B.1) provided by NGA.

GEOINT Analyst (Aeronautical)—acquire, analyze, and evaluate source and imagery in order to produce aeronautical products and mission-specific data in support of safety of navigation and national security requirements. They ensure the quality, accuracy, and currency of aeronautical information produced in-house, by contractors, and by national and international co-producers. They ensure the quality of aeronautical databases and products in accordance with ISO 9001 certification requirements.

GEOINT Analyst (Aeronautical Intelligence)—analyze and exploit worldwide aeronautical source and imagery in support of safety of navigation and national intelligence goals and requirements. They collaborate with partners and co-producers across the Department of Defense and the intelligence community to ensure that current navigation and intelligence information is available in databases and products. These analysts apply international certification standards to ensure the quality, accuracy, and currency of products and information.

GEOINT Analyst (Analytic Methodologist)—apply mathematical techniques for spatiotemporal analysis to solve complex military and intelligence problems in support of national security. They use analytic tools and techniques such as geographic information systems (GIS), quantitative methods and data visualization, modeling, systems analysis, comparative analysis, and database development. They provide technical input into the development, evaluation, use, and deployment of solutions and improvements to optimize GEOINT analysis and production. They also educate management and analysts in quantitative methods as they apply to GEOINT analysis.

GEOINT Analyst (Bathymetry)—receive, analyze, and deconflict U.S. and foreign bathymetric data (both digital and analog) for use by the intelligence community and external customers. They populate and maintain data and metadata in the bathymetry database, ensuring the accuracy of metadata pertaining to collection source platforms and depth recording devices.

GEOINT Analyst (Cartography)—acquire, analyze, compile, evaluate, and review geospatial data for use in standard products, nonstandard products, and/or data holdings in support of mission requirements. They attribute, exploit, extract, format, manipulate, position, and symbolize geospatial information. They ensure the quality, accuracy, and currency of geospatial information produced in-house or in cooperation with contractors and national and international co-producers for national, military, and civil partners.

GEOINT Analyst (Foundation Strategies)—collaborate with customers and source providers to manage tasking, collection, dissemination, and reporting related functions. They utilize unique systems and processes to support all mapping, charting, and geodesy global programs and products and operational and strategic requirements. They develop and coordinate tailored strategies within the intelligence community, create and adjudicate tasking and dissemination requirements for diverse customers, analyze and investigate collection performance for mapping, charting, and geodesy-specific data from commercial, national technical means, and other sources, and advise customers in support of the National System for Geospatial-Intelligence.

GEOINT Analyst (Geodetic Earth Sciences)—analyze the Earth's gravity and magnetic fields, geophysical structure, material properties, and dynamics for geospatial intelligence and Department of Defense applications. They define and maintain the World Geodetic System (WGS), perform datum transformations between WGS 84 and local datums, and develop spatial and temporal models defining Earth systems. They provide in-depth technical expertise on geodetic and geophysical issues to internal and external customers and represent NGA in external community forums establishing Department of Defense and intelligence community doctrine and policy.

GEOINT Analyst (Geodetic Orbit Sciences)—apply orbital mechanics and network systems analysis to acquire, process, and analyze satellite data. They develop, evaluate, and use algorithms and tools to provide integrity monitoring and determine precise ephemerides and other information. They analyze satellite ranging and timing information in order to identify anomalies in both real-time and post-fit scenarios. These analysts handle technical and logistical details pertinent to remote command and control of a worldwide network of satellite tracking stations. They provide in-depth technical expertise on satellite issues to internal and external customers, and represent NGA in external community forums establishing Department of Defense and intelligence community doctrine and policy.

GEOINT Analyst (Geodetic Survey)—solve three-dimensional geodetic and astronomic positional problems, determine geodetic and astronomic azimuths, and measure fluctuations and accelerations in the Earth's magnetic and gravity fields. They serve as NGA's primary ground-based data collectors and use and maintain a variety of geodetic and geophysical survey equipment to support data acquisition. They compute, adjust, and evaluate geodetic survey data acquired by other organizations. They provide technical expertise on geodetic and geophysical issues to customers and represent NGA in external community forums that establish Department of Defense and intelligence community doctrine and policy.

GEOINT Analyst (Geospatial Analysis)—produce intelligence products using geospatial methodologies and spatiotemporal data derived from imagery, intelligence databases, and other sources in support of national security. They are intelligence analysts who use their understanding of geographic information science and technology, spatial thinking, remote sensing, GIS, intelligence issues, and the social and physical sciences to create information, characterize events, and discover relationships and trends. They produce descriptive and predictive analyses and communicate findings as written, visual, and/or oral geospatial intelligence.

GEOINT Analyst (Geospatial Data Stewardship)—manage the generation, approval, and population of data within NGA's geospatial databases, ensuring that they can be used to satisfy customer requirements. They have read/write access to relevant databases, and make qualitative decisions about the data. They understand the capabilities and limitations of relevant database systems. They have a wide range of product/ data knowledge and understand the capabilities of the relevant systems to support the extraction, analysis, and finishing of in-house, co-production, commodity, and contract data.

GEOINT Analyst (Imagery Intelligence)—task and exploit imagery of all types in support of national and military security goals, concerns, and strategies. They conduct multi-INT research, populate intelligence databases, and produce written, graphic, and oral intelligence products. These analysts primarily analyze military force structure, capabilities, intentions, and vulnerabilities of adversaries and potential adversaries, weapons proliferation, emerging technologies, and treaty monitoring. They also work on diverse issues such as environmental concerns, counternarcotics, disaster assessments, infrastructure, underground facilities, and counterterrorism.

GEOINT Analyst (Imagery Science)—apply advanced techniques to determine the intelligence and geospatial information contained in imagery. They develop algorithms, evaluate tools, and create customized methodologies and products to address a variety of geospatial intelligence problems. They typically specialize in precision mensuration, radar, spectral, infrared, and other specialized collection systems.

GEOINT Analyst (Infrared Source Analysis)—specialize in metric reconstruction, trajectory analysis, and assessment of activity using nonliteral, infrared data from persistent geospatial intelligence sources to support characterization of foreign weapon system performance, counterproliferation, battlespace awareness, and other intelligence issues. The analysts apply their understanding of weapon systems, spatial and temporal reconstruction, error analysis, signature interpretation, sensor phenomenology, and orbital mechanics to help solve intelligence issues in accordance with the national intelligence priority framework. They communicate the meaning and significance of their analysis as written, visual, and/or oral geospatial intelligence.

GEOINT Analyst (Lidar Image Science)—analyze point cloud data to derive intelligence and geospatial information. They conduct feature extraction, characterize partially obscured objects, and perform change detection to support or extend analysis. They develop algorithms, evaluate tools, and create customized lidar methodologies to address a variety of geospatial intelligence problems.

GEOINT Analyst (Maritime)—acquire, analyze, compile, and disseminate maritime safety information and intelligence to populate and update nautical databases to support the digital nautical chart, hardcopy charts, digital publications, and electronic chart display and information systems. They generate mission-specific data and products, promulgate worldwide navigational warnings, and respond to queries from foreign hydrographic offices and the users of NGA products

and services. These duties are performed in fulfillment of the agency's marine navigation obligations under U.S. Code, Title 10, other federal and international laws and regulations, in support of national security goals, concerns, and strategies.

GEOINT Analyst (Nautical Cartography)—acquire, analyze, evaluate, and compile nautical products and mission-specific data in support of mission requirements and navigation safety. They evaluate information from a variety of sources to include foreign nautical charts, foreign notice to mariners, imagery, bathymetry, publications, ship reports, and other forms of geospatial intelligence against current NGA hydrographic data holdings. They ensure the quality, accuracy, and currency of nautical information produced either in-house or in cooperation with contractors and national and international co-producers for national, military, and civil customers.

GEOINT Analyst (Open Source Research)—discover, retrieve, and analyze open and classified source data and apply expertise in evaluating and acquiring sources of data. They develop regional and subject expertise in order to provide information tailored to mission requirements. They collaborate with other intelligence community agencies and outside institutions to enhance information access and resource sharing. They recommend and acquire source materials to enhance resource center collections and provide training to help customers optimize the use of open source in support of the GEOINT mission.

GEOINT Analyst (Persistent Operations)—specialize in the real-time execution of tasking, collection, processing, exploitation, and dissemination of persistent GEOINT resources and data. Real-time execution incorporates multi-INT collaborative inputs to address long-term and emerging customer requirements. Emphasis is placed on time-dominant assessment and reporting, sensor allocation, and responsive collection capabilities of persistent GEOINT sensors.

GEOINT Analyst (Photogrammetric Image Science)—apply advanced techniques to measure the precise dimensions or relative size of objects on imagery. This includes monoscopic, stereoscopic, overhead, handheld, or video imagery. They develop mensuration strategies, determine requirements, evaluate tools, and create customized methodologies and products to address a variety of geospatial intelligence problems.

GEOINT Analyst (Photogrammetry)—perform image assessment, point selection, mensuration, triangulation, orthorectification, and processing in order to produce standard and nonstandard image products. They increase the relative and absolute positioning accuracy of imagery from a variety of sensors to support mono and stereo exploitation. These analysts assess, evaluate, and extract elevation data from stereo imagery. They support both internal and external customers, review products produced by contractors and co-producers, and ensure that final products can be generated to meet customer's stringent accuracy requirements.

GEOINT Analyst (Political Geography)—apply expertise in foreign languages and Romanization policies to research and analyze sources containing foreign geographic names information, make policy recommendations for the standardization of foreign geographic names, and populate and maintain the Geographic Names Database. They monitor their area of interest for administrative, political, and infrastructure changes to maintain professional currency. They advise the U.S. Board of Geographic Names and serve as staff members of the Secretariat for the U.S. Board of Geographic Names Foreign Names Committee.

GEOINT Analyst (Radar Image Science)—apply advanced techniques to determine the intelligence and geospatial information contained in radar data. They develop algorithms, evaluate tools, and create customized radar methodologies and products to address a variety of geospatial intelligence problems.

GEOINT Analyst (Regional Geography)—apply knowledge and expertise in physical, sociocultural, and political aspects of countries, regions, and urban areas to support national security goals, concerns, and strategies. They gather and evaluate regional source data to build geospatial data layers that serve as a foundation for analytic work. Analyzing these data, they characterize events, discover relationships and trends, infer conclusions, predict behaviors, and communicate these

results as multisource geospatial intelligence in support of the intelligence community and the National System for Geospatial Intelligence.

GEOINT Analyst (Regional Source)—gather, assess, and evaluate source materials and ensure its quality and suitability to build foundational data. They conduct feasibility studies for the production of standard and specialty products. They maintain data integrity for various databases.

GEOINT Analyst (Requirements and Integration)—support the requirements, processes, and integration of future systems and methodologies. They coordinate with mission partners and customers to define requirements, evaluate, integrate, and transition to operations new technologies and associated functions.

GEOINT Analyst (Scientific Linguistics)—apply the principles of linguistics to problems encountered in the analysis and standardization of transliterated foreign geographic names. They collaborate with language experts at the national and international level to develop and implement consistent standardization policies. These analysts actively research country- and language-specific place-name issues and develop standardization policy recommendations for review and acceptance by the U.S. Board on Geographic Names Foreign Names Committee. They ensure the linguistic integrity of place-name intelligence developed by NGA on behalf of the U.S. Board on Geographic Names and the intelligence community.

GEOINT Analyst (Source Strategies)—collaborate with customers and source providers to develop comprehensive multi-INT, multisource strategies to address intelligence problems. They create tasking and dissemination requirements, adjudicate requirements, analyze and investigate collection performance, assess and report on end-to-end GEOINT system performance data, and advise customers in support of the National System for Geospatial Intelligence.

GEOINT Analyst (Spectral Image Science)—apply advanced techniques to determine the intelligence and geospatial information contained in the electro-optical region of the spectrum. This includes the exploitation of multi- and hyperspectral imagery. They develop algorithms, evaluate tools, and create customized spectral methodologies and products to address a variety of geospatial intelligence problems.

GEOINT Analyst (Thermal Infrared Image Science)—apply advanced techniques to determine the intelligence and geospatial information contained in thermal infrared imagery. They conduct phenomenological studies on objects or events of interest to inform intelligence conclusions. They develop algorithms, evaluate tools, and create customized infrared methodologies and products to address a variety of geospatial intelligence problems.

GEOINT Analyst (Throughput Strategies)—collaborate with customers and source providers to ensure the delivery of GEOINT source data. They develop optimal source production and dissemination requirements, monitor and report on network health and status, implement relevant release policies, monitor site bandwidth utilization, recommend alternative delivery paths, maintain user site profiles/allocation, and support operations integration of new data sources and system capabilities.

Principal Scientist—domain experts in a core agency scientific discipline. They define and lead scientific research strategy in their domain for NGA and the intelligence community. They apply extensive Department of Defense and intelligence community experience to identify and promote ideas and opportunities to advance the agency's ability to meet ever-expanding customer requirements for geospatial intelligence information. Principal Scientists act as a facilitator to bring diverse domains together to develop solutions based on an integration and fusion of technologies and sources. They support agency-level strategic planning. They provide oversight and coordination for agency science and technology programs.

Project Scientist—responsible for the day-to-day execution and technical oversight of a variety of scientific activities. They develop project schedules, determine resource requirements, provide technical guidance and oversight, and report results. Project Scientists apply in-depth expertise from a variety of scientific disciplines (e.g., photogrammetry, geodesy, GIS, computer science, mathematics, image science) to develop, analyze, evaluate, and apply new technology; develop expertise and tradecraft for the agency; and advise senior management on new and evolving technology. They participate in strategic planning, propose and defend program plans, and communicate and market results to customers and decision makers.

TABLE B.1 Education and Experience Requirements of Selected NGA Scientist and Analyst Positions

Education	Education and Experience	Experience
GEOINT Analyst (Aeronautical Intelligence)		
Bachelor's degree in aeronautical science, commercial aviation, flight, flight education, flight science, professional flight, professional flight technology, air traffic control, air traffic management, or another related degree that includes actual or simulated experience mandated with the curriculum along with a thorough knowledge of domestic and international air navigation principles as well as airspace and airfield infrastructure and operations	A minimum of 30 semester (45 quarter) hours of completed coursework in aeronautics, aerospace engineering, aerospace studies, geospatial information systems, global security and intelligence, cartography, homeland security, or other areas related to aeronautical navigation and operations. PLUS Minimum of 250 hours of flight experience as a pilot, co-pilot, navigator, or flight instructor OR Minimum of 3 years of civilian or military work experience as an air traffic controller, flight dispatcher, flight or ground school instructor, mission planner, aeronautical information specialist, terminal enroute procedural specialist, or other field which provided an understanding of air navigation principles, operations, publications, and airspace and airfield infrastructure and operating procedures	A minimum of 6 years of work experience within a NGA analytical occupation that involved the acquisition, collection, analysis and evaluation, extraction and population, and maintenance of NGA geospatial or safety of navigation related databases
GEOINT Analyst (Analytic Methodologist)		
Bachelor's degree in applied mathematics, geographic information science, geography, physical science, operations research, statistics, or a related discipline	Four years, including a minimum of 24 semester (36 quarter) hours of coursework in any area listed in the education requirements and an additional 6 semester (9 quarter) hours of college-level nonbusiness mathematics or statistics (e.g., college algebra, trigonometry, calculus, inferential statistics) plus experience working as an intelligence analyst or in a closely related field that demonstrates the ability to successfully perform the tasks associated with this work	
GEOINT Analyst (Bathymetry)		
Bachelor's degree in geography, geology, hydrography, hydrology, marine sciences, oceanography, physical science, remote sensing, or a related discipline, or a bachelor's degree with 30 semester hours of coursework in the above disciplines. Designation as an American Congress on Surveying and Mapping-The Hydrographic Society of America Certified Hydrographer is highly desired	Four years, including a minimum of 30 semester (45 quarter) hours of coursework in any area listed in the education requirements, plus experience that demonstrates the ability to successfully perform the duties associated with this work	Six years of experience in the disciplines of hydrography or bathymetry that includes marine surveying, the use of current GIS tools, methods of research and analysis, application of hydrographic or bathymetric principles, or work related to the disciplines listed in the education requirements

continued

TABLE B.1 Continued

Education	Education and Experience	Experience
GEOINT Analyst (Cartography)		
Bachelor's degree in cartography or a major that included, or was supplemented by, at least 30 semester (45 quarter) hours of coursework in cartography and/or directly related sciences and mathematics. Coursework may include, but is not limited to, astronomy, cartography, forestry, geodesy, geology, geophysics, GIS, land surveying, photogrammetry, physical and geological oceanography, geography, and remote sensing. Computer software classes associated with current technology, GIS courses, and information management classes may also be counted. The 30 semester (45 quarter) hours must include at least 6 semester (9 quarter) hours of college-level nonbusiness mathematics or statistics, but these should not account for more than 15 semester (22 quarter) hours	Four years, including a minimum of 30 semester (45 quarter) hours of coursework in any area listed in the education requirements plus experience that demonstrates the ability to successfully perform the duties associated with this work	Six years of significant applied experience in the cartographic field that includes the use of current GIS tools
GEOINT Analyst (Foundation Strategies)		
Bachelor's degree in cartography, cultural area studies, earth sciences, environmental science, geodesy, geography, geology, GIS, history, hydrography, hydrology, imagery science, international affairs/studies, liberal studies, oceanography, photogrammetry, physical science, political science, remote sensing, or a related discipline	Four years, including a minimum of 24 semester (36 quarter) hours of coursework in any area listed in the education requirements plus experience that demonstrates the ability to successfully perform the duties associated with this work	Six years of significant applied experience in the intelligence field, imagery analysis, cartography, the use of current GIS tools, or work related to the fields listed in the education requirements
GEOINT Analyst (Geodetic Earth Sciences)		
Bachelor's degree in geodesy, mathematics, physical science, or a related discipline that includes at least 30 semester (45 quarter) hours of coursework in any combination of astronomy, cartography, computer science, engineering science, geodesy, geology, geomatics, geophysics, mathematics, meteorology, orbital mechanics, photogrammetry, physical science, physics, remote sensing, or surveying. Coursework must include differential equations and integral calculus	Four years, including a minimum of 30 semester (45 quarter) hours of coursework in any area listed in the education requirements plus experience that demonstrates the ability to successfully perform the duties associated with this work	Six years of experience in conducting work related to civil engineering, geodesy, geophysics, geotechnical analysis, geodetic surveying, or related experience
GEOINT Analyst (Geodetic Orbit Sciences)		
Bachelor's degree in geodesy, mathematics, physical science, or a related discipline that includes at least 30 semester (45 quarter) hours of coursework in any combination of astronomy, cartography, computer science, engineering science, geodesy, geology, geomatics, geophysics, mathematics, meteorology, orbital mechanics, photogrammetry, physical science, physics, remote sensing, or surveying. Coursework must include differential equations and integral calculus	Four years, including a minimum of 30 semester (45 quarter) hours of coursework in any area listed in the education requirements plus experience that demonstrates the ability to successfully perform the duties associated with this work	Six years of experience in conducting work related to satellite operations, GPS surveying, geodesy, geophysics, or Wide Area Network analysis

continued

TABLE B.1 Continued

Education	Education and Experience	Experience
GEOINT Analyst (Geodetic Survey)		
Bachelor's degree in geodesy, mathematics, physical science, or a related discipline that includes at least 30 semester (45 quarter) hours of coursework in any combination of astronomy, cartography, computer science, engineering science, geodesy, geology, geomatics, geophysics, GIS, mathematics, meteorology, orbital mechanics, photogrammetry, physical science, physics, remote sensing, or surveying. Coursework must include differential equations and integral calculus	Four years, including a minimum of 30 semester (45 quarter) hours of coursework in any area listed in the education requirements plus experience that demonstrates the ability to successfully perform the duties associated with this work	Six years of experience in conducting work related to civil engineering, geodesy, geophysics, geotechnical analysis, surveying, or related experience. Classification as a Professional Engineer or Land Surveyor is highly desirable
GEOINT Analyst (Geospatial Analysis)		
Bachelor's degree in a cartography, geography, GIS, physical science, applied mathematics, statistics, or a related discipline, or a bachelor's degree in any discipline with a certificate in GIS from an accredited university	Four years, including a minimum of 24 semester (36 quarter) hours of coursework in GIS, spatial analysis, or any area listed in the education requirements plus additional experience that demonstrates the ability to successfully perform the duties associated with this work. A certificate in GIS from an accredited university or an emphasis in GIS is highly desired	Six years of significant applied experience in a geospatial analysis field that includes the use of current GIS tools, methods of research and analysis, application of cartographic principles, and work related to the fields listed in the education requirements
GEOINT Analyst (Geospatial Data Stewardship)		
Bachelor's degree in cartography, computer science, geology, geography, geomatics, GIS, physical science, urban and regional planning, or a related discipline. The 30 semester (45 quarter) hours must include at least 6 semester (9 quarter) hours of college-level nonbusiness mathematics or statistics, but these should not account for more than 15 semester (22 quarter) hours	Four years, including a minimum of 30 semester (45 quarter) hours of coursework in any area listed in the education requirements plus experience that demonstrates the ability to successfully perform the duties associated with this work	Six years of significant applied experience in a field that includes geospatial analysis, feature extraction, and data evaluation and manipulation
GEOINT Analyst (Imagery Intelligence)		
Bachelor's degree in engineering, foreign area studies, geography, history, imagery science, international affairs, military science, physical science, political science, remote sensing, or a related discipline	Four years, including a minimum of 24 semester (36 quarter) hours of college coursework in any area listed in the education requirements plus experience working as an imagery analyst, OR in a closely related field that demonstrates the ability to successfully perform the tasks associated with this work	Six years of significant applied experience in imagery analysis or all-source intelligence experience, or six years in combat arms or in combat support for military personnel
GEOINT Analyst (Imagery Science)		
Bachelor's degree in engineering, imagery science, mathematics, physical science, or a related discipline	Four years, including a minimum of 24 semester (36 quarter) hours of coursework in any area listed in the education requirements, plus experience that demonstrates the ability to successfully perform the duties associated with this work	
GEOINT Analyst (Infrared Source Analysis)		
Bachelor's degree in mathematics, engineering, geography, physical science, remote sensing, or a related discipline	Four years, including a minimum of 24 semester (36 quarter) hours of college coursework in any area listed in the education requirements, plus experience working in a geospatial intelligence discipline or in a closely related field that demonstrates the ability to successfully perform the tasks associated with this work	A minimum of 4 years of relevant military experience or experience working in a geospatial intelligence discipline

continued

TABLE B.1 Continued

Education	Education and Experience	Experience
GEOINT Analyst (Lidar Image Science)		
Bachelor's degree in engineering, imagery science, mathematics, physical science, or a related discipline	Four years, including a minimum of 24 semester (36 quarter) hours of coursework in any area listed in the education requirements, plus experience that demonstrates the ability to successfully perform the duties associated with this work	
GEOINT Analyst (Maritime)		
Bachelor's degree from a federal or state maritime academy, the U.S. Naval Academy, or the U.S. Coast Guard Academy. Education must include successful completion of the requirements for a U.S. Coast Guard Third Mate's License or Officer of the Deck Underway Qualification by the U.S. Navy or U.S. Coast Guard. Strongly desired candidates will also include classes in GIS software		Three years of general experience that demonstrates knowledge in the field of marine navigation. Examples of qualifying experience include navigation and sea experience on U.S. oceangoing vessels, experience as an instructor on marine navigational subjects, or other related experience. Experience must include successful completion of the requirements for a U.S. Coast Guard Third Mate's License or Officer of the Deck Underway Qualification by the U.S. Navy or U.S. Coast Guard. Strongly desired candidates will also include classes in GIS software
GEOINT Analyst (Open Source Research)		
A Master of Science in library science, library and information science, information science and learning technologies, or a similar program from an accredited university OR a bachelor's degree in cartography, cultural area studies, earth sciences, engineering, environmental science, foreign language, geodesy, geography, GIS, history, hydrography, hydrology, imagery science, international affairs/studies, oceanography, photogrammetry, physical science, remote sensing, or a related discipline	Four years, including a minimum of 24 semester (36 quarter) hours of coursework in any area listed in the bachelor's education requirements plus experience that demonstrates the ability to successfully perform the duties associated with this work	Six years of significant applied experience in information retrieval that includes methods of research and analysis, the use of current GIS tools, or work related to the fields listed in the bachelor's education requirements
GEOINT Analyst (Persistent Operations)		
Bachelor's degree in science, technology, business administration, or a related field	Four years, including a minimum of 24 semester (36 quarter) hours of college coursework in any area listed in the education requirements plus experience working in a GEOINT related field that demonstrates the ability to successfully perform the tasks associated with this work	Six years of experience in a GEOINT-related field
GEOINT Analyst (Photogrammetric Image Science)		
Bachelor's degree in engineering, imagery science, mathematics, physical science, or a related discipline	Four years, including a minimum of 24 semester (36 quarter) hours of coursework in any area listed in the education requirements, plus experience that demonstrates the ability to successfully perform the duties associated with this work	

continued

TABLE B.1 Continued

Education	Education and Experience	Experience
GEOINT Analyst (Photogrammetry)		
Bachelor's degree in cartography or a major that included, or was supplemented by, at least 30 semester (45 quarter) hours of coursework in cartography and/or directly related sciences and mathematics. Coursework may include, but is not limited to, cartography, computer science, geodesy, geology, geophysics, land surveying, photogrammetry, geography, remote sensing, or a related discipline. The 30 semester (45 quarter) hours must include at least 6 semester (9 quarter) hours of college-level nonbusiness mathematics or statistics, but these should not account for more than 15 semester (22 quarter) hours	Four years, including a minimum of 30 semester (45 quarter) hours of coursework in any area listed in the education requirements plus experience that demonstrates the ability to successfully perform the duties associated with this work	Six years of significant applied experience in the cartographic field that includes the application of mensuration techniques
GEOINT Analyst (Political Geography)		
Bachelor's degree in anthropology, cartography, cultural area studies, foreign language, geography, GIS, history, international affairs/studies, linguistics, or a related discipline	Four years, including a minimum of 24 semester (36 quarter) hours of coursework in any area listed in the education requirements, plus experience that demonstrates the ability to successfully perform the duties associated with this work	Six years of significant applied experience in geography that includes the application of foreign language or linguistics, methods of research and analysis, application of cartographic principles, and work related to the fields listed in the education requirements
GEOINT Analyst (Radar Image Science)		
Bachelor's degree in engineering, imagery science, mathematics, physical science, or a related discipline	Four years, including a minimum of 24 semester (36 quarter) hours of coursework in any area listed in the education requirements, plus experience that demonstrates the ability to successfully perform the duties associated with this work	
GEOINT Analyst (Regional Geography)		
Bachelor's degree in anthropology, cartography, cultural area studies, earth sciences, foreign language, geography, GIS, history, international affairs/studies, sociology, or a related discipline	Four years, including a minimum of 24 semester (36 quarter) hours of coursework in any area listed in the education requirements, plus experience that demonstrates the ability to successfully perform the duties associated with this work	Six years of significant applied experience in geography or related fields that includes the use of current GIS tools, methods of research and analysis, application of cartographic principles, and work related to the fields listed in the education requirements
GEOINT Analyst (Regional Source)		
Bachelor's degree in cartography, geography, GIS, information science, library science, or a related discipline	Four years, including a minimum of 24 semester (36 quarter) hours of coursework in any area listed in the education requirements, plus experience that demonstrates the ability to successfully perform the duties associated with this work	Six years of significant applied experience in the use of current GIS tools, methods of research and analysis, application of cartographic principles, and work related to the fields listed in the education requirements
GEOINT Analyst (Requirements & Integration)		
None	Six years of education and experience. Education and experience in remote sensing, geography, intelligence, or equivalent areas. Experience in acquisition and/or program management is highly desirable	Six years of significant applied experience in remote sensing, geography, intelligence, or equivalent areas

continued

TABLE B.1 Continued

Education	Education and Experience	Experience
GEOINT Analyst (Scientific Linguistics)		
Bachelor's degree in linguistics or an NGA-specified foreign language with a minimum of 6 semester (9 quarter) hours of coursework in linguistics. Master's degree in theoretical linguistics is highly desirable. The candidate must also receive a reading proficiency level 2 on the Defense Language Proficiency Test in a foreign language specified by NGA	Four years in a NGA-specified foreign language that includes a minimum of 6 semester (9 quarter) hours of coursework in linguistics and demonstrates the ability to successfully perform the duties associated with this work. The candidate must also receive a reading proficiency level 2 on the Defense Language Proficiency Test in a foreign language specified by NGA	
GEOINT Analyst (Source Strategies)		
Bachelor's degree in cartography, cultural area studies, earth sciences, environmental science, geodesy, geography, geology, GIS, history, hydrography, hydrology, imagery science, international affairs/studies, liberal studies, oceanography, photogrammetry, physical science, political science, remote sensing, or a related discipline	Four years, including a minimum of 24 semester (36 quarter) hours of coursework in any area listed in the education requirements plus experience that demonstrates the ability to successfully perform the duties associated with this work	Six years of significant applied experience in the intelligence field, imagery analysis, the use of current GIS tools, or work related to the fields listed in the education requirements
GEOINT Analyst (Spectral Image Science)		
Bachelor's degree in engineering, imagery science, mathematics, physical science, or a related discipline	Four years, including a minimum of 24 semester (36 quarter) hours of coursework in any area listed in the education requirements, plus experience that demonstrates the ability to successfully perform the duties associated with this work	
GEOINT Analyst (Thermal Infrared Image Science)		
Bachelor's degree in engineering, imagery science, mathematics, physical science, or a related discipline	Four years, including a minimum of 24 semester (36 quarter) hours of coursework in any area listed in the education requirements, plus experience that demonstrates the ability to successfully perform the duties associated with this work	
GEOINT Analyst (Throughput Strategies)		
Bachelor's degree in cartography, communications, cultural area studies, earth sciences, engineering, environmental science, geography, geology, GIS, history, imagery science, international affairs/ studies, liberal studies, photogrammetry, physical science, remote sensing, or a related discipline	Five years, including a minimum of 24 semester (36 quarter) hours of coursework in any area listed in the education requirements plus experience that demonstrates the ability to successfully perform the duties associated with this work	Six years of significant applied experience in the application of communication strategies, architecture, systems integration, throughput management, requirements analysis, or work related to the fields listed in the education requirements
Principal Scientist		
Bachelor's degree in engineering, mathematics, physical science, or a related discipline that includes at least 24 semester (36 quarter) hours in physical science and/or related engineering science. Such coursework includes, but is not limited to, astronomy, cartography, chemistry, computer science, dynamics, electrical engineering, geodesy, geology, geophysics, GIS, mathematics, orbital mechanics, photogrammetry, physics, remote sensing, or surveying. An advanced degree (e.g., M.S., Ph.D.) in engineering, mathematics, physical science, or a related discipline is preferred	Four years, including minimum of 24 semester (36 quarter) hours of coursework in any area listed in the education requirements plus additional experience that demonstrates the ability to successfully perform the duties associated with this work	

continued

TABLE B.1 Continued

Education	Education and Experience	Experience
Project Scientist		
Bachelor's degree in engineering, mathematics, physical science, or a related discipline that includes 24 semester (36 quarter) hours in physical science and/or a related engineering science. Such coursework includes, but is not limited to, astronomy, cartography, chemistry, computer science, dynamics, electrical engineering, geodesy, geology, geophysics, GIS, mathematics, orbital mechanics, photogrammetry, physics, remote sensing, or surveying. Although not mandatory, coursework in differential and integral calculus is preferred	Four years, including a minimum of 24 semester (36 quarter) hours of college education in any areas listed in the education requirements plus experience that demonstrates the ability to successfully perform the duties associated with this work	

NOTE: As a rule, every 30 semester (45 quarter) hours of college work is equivalent to one year of experience.
SOURCE: NGA.

Appendix C

Data on Instructional Programs and Citizenship

DEGREE DATA

TABLE C.1 Department of Education Codes and Descriptions of 164 Instructional Programs That Are Relevant to NGA

Code	Title	Description
03.0101	Natural resources / conservation, general	A general program that focuses on the studies and activities relating to the natural environment and its conservation, use, and improvement. Includes instruction in subjects such as climate, air, soil, water, land, fish and wildlife, and plant resources; in the basic principles of environmental science and natural resources management; and the recreational and economic uses of renewable and nonrenewable natural resources
03.0102	Environmental science / studies	(Deleted) Report under code 03.0103 or 03.0104
03.0103	Environmental studies	(NEW) A program that focuses on environment-related issues using scientific, social scientific, or humanistic approaches or a combination. Includes instruction in the basic principles of ecology and environmental science and related subjects such as policy, politics, law, economics, social aspects, planning, pollution control, natural resources, and the interactions of human beings and nature
03.0104	Environmental science	(NEW) A program that focuses on the application of biological, chemical, and physical principles to the study of the physical environment and the solution of environmental problems, including subjects such as abating or controlling environmental pollution and degradation; the interaction between human society and the natural environment; and natural resources management. Includes instruction in biology, chemistry, physics, geosciences, climatology, statistics, and mathematical modeling
03.0204	Natural resource economics	(NEW) A program that focuses on the application of economic concepts and methods to the analysis of issues such as air and water pollution, land use planning, waste disposal, invasive species and pest control, conservation policies, and related environmental problems. Includes instruction in cost-benefit analysis; environmental impact assessment; evaluation and assessment of alternative resource management strategies; policy evaluation and monitoring; and descriptive and analytic tools for studying how environmental developments affect the economic system
03.0205	Water, wetlands, and marine resources management	(NEW) A program that prepares individuals to apply the principles of marine/aquatic biology, oceanography, natural resource economics, and natural resources management to the development, conservation, and management of freshwater and saltwater environments. Includes instruction in subjects such as wetlands, riverine, lacustrin, coastal, and oceanic water resources; water conservation and use; flood control; pollution control; water supply logistics; wastewater management; aquatic and marine ecology; aquatic and marine life conservation; and the economic and recreational uses of water resources
03.0206	Land use planning and management / development	(NEW) A program that focuses on how public and/or private land and associated resources can be preserved, developed, and used for maximum social, economic, and environmental benefit. Includes instruction in natural resources management, natural resource economics, public policy, regional and land use planning, environmental impact assessment, applicable law and regulations, government and politics, principles of business and real estate land use, statistical and analytical tools, computer applications, mapping and report preparation, site analysis, cost analysis, and communications skills

continued

TABLE C.1 Continued

Code	Title	Description
03.0501	Forestry, general	A program that generally prepares individuals to manage and develop forest areas for economic, recreational, and ecological purposes. Includes instruction in forest-related sciences, mapping, statistics, harvesting and production technology, natural resources management and economics, wildlife sciences, administration, and public relations
03.0502	Forest sciences and biology	A program that focuses on the application of one or more forest-related sciences to the study of environmental factors affecting forests and the growth and management of forest resources. Includes instruction in forest biology, forest hydrology, forest mensuration, silviculture, forest soils, water resources, environmental science, forest resources management, and wood science
03.0506	Forest management / forest resources management	A program that prepares individuals to apply principles of forestry and natural resources management to the administration of forest lands and related resources. Includes instruction in silviculture, forest mensuration, forest protection, inventorying, biometrics, geographic information systems, remote sensing, photogrammetry, forest policy and economics, forest land use planning, fire protection and management, and related administrative skills
03.0508	Urban forestry	(NEW) A program that prepares individuals to apply the principles of forestry and related sciences to the development, care, and maintenance of individual trees and forested areas within or close to areas of dense human habitation. Includes instruction in urban environments; effects of pollution on tree species; environmental design and landscaping; urban pest infestation; urban forest management; and applicable policies and regulations
03.0509	Wood science and wood products / pulp and paper technology	A program that focuses on the application of chemical, physical, and engineering principles to the analysis of the properties and behavior of wood and wood products and the development of processes for converting wood into paper and other products. Includes instruction in wood classification and testing, product development, manufacturing and processing technologies, and the design and development of related equipment and systems
04.0301	City / urban, community and regional planning	A program that prepares individuals to apply principles of planning, analysis, and architecture to the development and improvement of urban areas and surrounding regions, and to function as professional planners. Includes instruction in principles of architecture; master plan development; service, communications, and transportation systems design; community and commercial development; zoning; land use planning; applied economics; policy analysis; applicable laws and regulations; and professional responsibilities and managerial duties
05.0101	African studies	A program that focuses on the history, society, politics, culture, and economics of one or more of the peoples of the African Continent, usually with an emphasis on Africa south of the Sahara, and including the African diaspora overseas
05.0102	American / United States studies / civilization	A program that focuses on the history, society, politics, culture, and economics of the United States and its Pre-Columbian and colonial predecessors, and including the flow of immigrants from other societies
05.0103	Asian studies / civilization	A program that focuses on the history, society, politics, culture, and economics of one or more of the peoples of the Asian Continent, including the study of the Asian diasporas overseas
05.0104	East Asian studies	A program that focuses on the history, society, politics, culture, and economics of one or more of the peoples of East Asia, defined as including China, Korea, Japan, Mongolia, Taiwan, Tibet, related borderlands and island groups, and including the study of the East Asian diasporas overseas
05.0105	Central / middle and eastern European studies	A program that focuses on the history, society, politics, culture, and economics of one or more of the peoples of what is historically known as Central/Middle and Eastern Europe, defined as including Austria, the Balkans, the Baltic States, Belarus, Czech Republic, Hungary, Romania, Poland, Russia, Slovakia, Ukraine, related borderlands and island groups, and migration patterns
05.0106	European studies / civilization	A program that focuses on the history, society, politics, culture, and economics of one or more of the peoples of the European Continent, including the study of European migration patterns and colonial empires
05.0107	Latin American studies	A program that focuses on the history, society, politics, culture, and economics of one or more of the Hispanic peoples of the North and South American Continents outside Canada and the United States, including the study of the Pre-Columbian period and the flow of immigrants from other societies
05.0108	Near and Middle Eastern studies	A program that focuses on the history, society, politics, culture, and economics of one or more of the peoples of North Africa, Southwestern Asia, Asia Minor, and the Arabian Peninsula, related borderlands and island groups, and including emigrant and immigrant groups
05.0109	Pacific area / Pacific Rim studies	A program that focuses on the history, society, politics, culture, and economics of one or more of the peoples of Australasia and the Pacific Ocean, related island groups and bordering coastal regions, and including pre- and post-colonial migration patterns

continued

TABLE C.1 Continued

Code	Title	Description
05.0110	Russian studies	A program that focuses on the history, society, politics, culture, and economics of one or more of the peoples of the Russian Federation and its Soviet, Czarist, and medieval predecessors and related borderlands
05.0111	Scandinavian studies	A program that focuses on the history, society, politics, culture, and economics of one or more of the peoples of Scandinavia, defined as Northern Europe including Denmark, Finland, Iceland, Norway, Sweden, related island groups (including Greenland), and borderlands
05.0112	South Asian studies	A program that focuses on the history, society, politics, culture, and economics of one or more of the peoples of South Asia, defined as including Afghanistan, India, the Maldives, Myanmar (Burma), Pakistan, and Sri Lanka and related borderlands and island groups; and including the study of migration patterns and overseas diasporas
05.0113	Southeast Asian studies	A program that focuses on the history, society, politics, culture, and economics of one or more of the peoples of Southeast Asia, defined as including Brunei, Cambodia, Indonesia, Laos, Malaysia, The Philippines, Singapore, Thailand, and Viet Nam; related borderlands and island groups; and including the study of migration patterns and overseas diasporas
05.0114	Western European studies	A program that focuses on the history, society, politics, culture, and economics of one or more of the peoples of historical Western Europe, defined as including Britain, Ireland, France, the Low Countries, the Iberian Peninsula, Italy, the Western Mediterranean, and related island groups and borderlands
05.0115	Canadian studies	A program that focuses on the history, society, politics, culture, and economics of one or more of the peoples of Canada and its Pre-Columbian, colonial, and pre-federation predecessors, including immigrant flows and related borderlands and island groups
05.0116	Balkans studies	(NEW) A program that focuses on the history, society, politics, culture, and economics of one or more of the peoples inhabiting the Balkan Peninsula and associated island groups and borderlands, Southern Slavic and non-Slavic, during the medieval, Ottoman, and modern periods
05.0117	Baltic studies	(NEW) A program that focuses on the history, society, politics, culture, and economics of one or more of the peoples inhabiting the coastlands of the Baltic Sea, including Baltic, Germanic, Scandinavian, and Slavic populations, related borderlands and island groups, and problems of cultural survival and assimilation
05.0118	Slavic studies	(NEW) A program that focuses on the history, society, politics, culture, and economics of one or more of the Slavic peoples inhabiting Europe, Asia, and in immigrant groups elsewhere, including the study of the emergence and migration patterns of Slavic culture, languages, and populations
05.0119	Caribbean studies	(NEW) A program that focuses on the history, society, politics, culture, and economics of one or more of the peoples inhabiting the major islands and archipelagoes of the Caribbean Sea and related coastal borderlands, including immigration patterns and Pre-Columbian, colonial, and modern societies
05.0120	Ural-Altaic and central Asian studies	(NEW) A program that focuses on the history, society, politics, culture, and economics of one or more of the peoples and countries of Inner/Central Asia, including the Turkic and Mongolian inhabitants of the Caspian, Amur, Tien Shan, Baikal, Gobi, Siberian, and Manchurian areas and the historical Silk Road, in terms of their past and present development
05.0121	Commonwealth studies	(NEW) A program that focuses on the history, society, politics, culture, and economics of one or more of the peoples and countries comprising components of the historical British Empire and modern British Commonwealth, including migration patterns, shared sociocultural and political features and problems, and contemporary relations
05.0122	Regional studies (U.S., Canadian, foreign)	(NEW) A program that focuses on the defined geographic subregions and subcultures within modern and historical countries and societies. Includes such topics as Acadian studies, French Canadian and Quebec studies, Southern (U.S.) studies, Appalachian (U.S.) studies, New England studies, Southwestern studies, Northern studies, and others
05.0123	Chinese studies	(NEW) A program that focuses on the history, society, politics, culture, and economics of one or more of the peoples of present-day China and its historical predecessors, related borderlands and island groups, and the overseas Chinese diaspora
05.0124	French studies	(NEW) A program that focuses on the history, society, politics, culture, and economics of France, other Francophone countries inside and outside Europe, and the French colonial experience and the associated French minorities around the world
05.0125	German studies	(NEW) A program that focuses on the history, society, politics, culture, and economics of Germany, the neighboring countries of Austria and Switzerland, the German minorities in neighboring European countries, and the historical areas of German influence across Europe and overseas

continued

TABLE C.1 Continued

Code	Title	Description
05.0126	Italian studies	(NEW) A program that focuses on the history, society, politics, culture, and economics of modern Italy and its predecessors on the Italian Peninsula, including overseas migrations of Italian peoples
05.0127	Japanese studies	(NEW) A program that focuses on the history, society, politics, culture, and economics of the peoples of Japan, and related island groups and coastal neighbors
05.0128	Korean studies	(NEW) A program that focuses on the history, society, politics, culture, and economics of the peoples of Korea, including related island groups and borderlands
05.0129	Polish studies	(NEW) A program that focuses on the history, society, politics, culture, and economics of Poland and the current and historical inhabitants of the Polish lands, including borderlands, from earliest times to the present
05.0130	Spanish and Iberian studies	(NEW) A program that focuses on the history, society, politics, culture, and economics of the peoples of the Iberian Peninsula and related island groups and border regions from earliest times to the present, with particular emphasis on the development of Spain and Portugal but including other historical and current cultures
05.0131	Tibetan studies	(NEW) A program that focuses on the history, society, politics, culture, and economics of Tibet and its borderlands, with emphasis on both pre-modern and modern Tibet and associated religious and exile movements
05.0132	Ukraine studies	(NEW) A program that focuses on the history, society, politics, culture, and economics of Ukraine and its inhabitants, and related border regions, from earliest times to the present
05.0199	Area studies, other	Any instructional program in specifically defined area studies not listed above
10.0301	Graphic communications, general	(NEW) A program that generally prepares individuals to apply technical knowledge and skills in the manufacture and distribution or transmission of graphic communications products. Includes instruction in the prepress, press, and postpress phases of production operations and processes such as offset lithography, flexography, gravure, letterpress, screen printing, foil stamping, digital imaging, and other reproduction methods
10.0302	Printing management	(NEW) A program that prepares individuals to apply technical and managerial knowledge and skills to the processes and procedures of managing printing operations from initial design through finished product distribution. Includes instruction in the principles of graphic communications design and production; quality control; printing operations management; computerization; printing plant management; business finance and marketing; logistics and distribution; personnel supervision and leadership; and professional standards in the graphic communications industry
10.0303	Prepress / desktop publishing and digital imaging design	A program that prepares individuals to apply technical knowledge and skills to the layout, design and typographic arrangement of printed and/or electronic graphic and textual products. Includes instruction in printing and lithographic equipment and operations; computer hardware and software; digital imaging; print preparation; page layout and design; desktop publishing; and applicable principles of graphic design and web page design
10.0304	Animation, interactive technology, video graphics and special effects	(NEW) A program that prepares individuals to use computer applications and related visual and sound imaging techniques to manipulate images and information originating as film, video, still photographs, digital copy, soundtracks, and physical objects in order to communicate messages simulating real-world content. Includes instruction in specialized camerawork and equipment operation and maintenance, image capture, computer programming, dubbing, CAD applications, and applications to specific commercial, industrial, and entertainment needs
10.0308	Computer typography and composition equipment operator	A program that prepares individuals to apply technical knowledge and skills to design and execute page formats, layouts and text composition, and to make typographical selections using computer graphics and other computer-assisted design programs
10.0399	Graphic communications, other	(NEW) Any instructional program in graphic communications not listed above
11.0101	Computer and information sciences, general	A general program that focuses on computing, computer science, and information science and systems as part of a broad and/or interdisciplinary program. Such programs are undifferentiated as to title and content and are not to be confused with specific programs in computer science, information science, or related support services
11.0102	Artificial intelligence and robotics	(NEW) A program that focuses on the symbolic inference, representation, and simulation by computers and software of human learning and reasoning processes and capabilities, and the modeling of human motor control and motions by computer-driven machinery. Includes instruction in computing theory, cybernetics, human factors, natural language processing, robot design, and applicable aspects of engineering, technology, and specific end-use applications

continued

TABLE C.1 Continued

Code	Title	Description
11.0103	Information technology	(NEW) A program that focuses on the design of technological information systems, including computing systems, as solutions to business and research data and communications support needs. Includes instruction in the principles of computer hardware and software components, algorithms, databases, telecommunications, user tactics, application testing, and human interface design
11.0199	Computer and information sciences, other	(NEW) Any instructional program in computer science not listed above
11.0201	Computer programming / programmer, general	A program that focuses on the general writing and implementation of generic and customized programs to drive operating systems and that generally prepares individuals to apply the methods and procedures of software design and programming to software installation and maintenance. Includes instruction in software design, low- and high-level languages and program writing; program customization and linking; prototype testing; troubleshooting; and related aspects of operating systems and networks
11.0202	Computer programming, specific applications	(NEW) A program that prepares individuals to apply the knowledge and skills of general computer programming to the solution of specific operational problems and customization requirements presented by individual software users and organizational users. Includes training in specific types of software and its installation and maintenance
11.0203	Computer programming, vendor / product certification	(NEW) A program that prepares individuals to fulfill the requirements set by vendors for professional qualification as certified installation, customization, and maintenance engineers for specific software products and/or processes. Includes training in specific vendor supported software products and their installation and maintenance
11.0299	Computer programming, other	(NEW) Any instructional program in computer programming not listed above
11.0301	Data processing and data processing technology / technician	A program that prepares individuals to master and use computer software programs and applications for inputting, verifying, organizing, storing, retrieving, transforming (changing, updating, and deleting), and extracting information. Includes instruction in using various operating system configurations and in types of data entry such as word processing, spreadsheets, calculators, management programs, design programs, database programs, and research programs
11.0401	Information science / studies	A program that focuses on the theory, organization, and process of information collection, transmission, and utilization in traditional and electronic forms. Includes instruction in information classification and organization; information storage and processing; transmission, transfer, and signaling; communications and networking; systems planning and design; human interfacing and use analysis; database development; information policy analysis; and related aspects of hardware, software, economics, social factors, and capacity
11.0501	Computer systems analysis / analyst	A program that prepares individuals to apply programming and systems analysis principles to the selection, implementation, and troubleshooting of customized computer and software installations across the life cycle. Includes instruction in computer hardware and software; compilation, composition, execution, and operating systems; low- and high-level languages and language programming; programming and debugging techniques; installation and maintenance testing and documentation; process and data flow analysis; user needs analysis and documentation; cost-benefit analysis; and specification design
11.0701	Computer science	A general program that focuses on computers, computing problems and solutions, and the design of computer systems and user interfaces from a scientific perspective. Includes instruction in the principles of computational science, and computing theory; computer hardware design; computer development and programming; and applications to a variety of end-use situations
11.0801	Web page, digital / multimedia and information resources design	(NEW) A program that prepares individuals to apply HTML, XML, Javascript, graphics applications, and other authoring tools to the design, editing, and publishing (launching) of documents, images, graphics, sound, and multimedia products on the World Wide Web. Includes instruction in Internet theory; web page standards and policies; elements of web page design; user interfaces; vector tools; special effects; interactive and multimedia components; search engines; navigation; morphing; e-commerce tools; and emerging web technologies
11.0802	Data modeling / warehousing and database administration	(NEW) A program that prepares individuals to design and manage the construction of databases and related software programs and applications, including the linking of individual data sets to create complex searchable databases (warehousing) and the use of analytical search tools (mining). Includes instruction in database theory, logic, and semantics; operational and warehouse modeling; dimensionality; attributes and hierarchies; data definition; technical architecture; access and security design; integration; formatting and extraction; data delivery; index design; implementation problems; planning and budgeting; and client and networking issues

continued

TABLE C.1 Continued

Code	Title	Description
11.0803	Computer graphics	(NEW) A program that focuses on the software, hardware, and mathematical tools used to represent, display, and manipulate topological, two-, and three-dimensional objects on a computer screen and that prepares individuals to function as computer graphics specialists. Includes instruction in graphics software and systems; digital multimedia; graphic design; graphics devices, processors, and standards; attributes and transformations; projections; surface identification and rendering; color theory and application; and applicable geometry and algorithms
11.0899	Computer software and media applications, other	(NEW) Any instructional program in computer software and media applications not listed above
11.0901	Computer systems networking and telecommunications	(NEW) A program that focuses on the design, implementation, and management of linked systems of computers, peripherals, and associated software to maximize efficiency and productivity, and that prepares individuals to function as network specialists and managers at various levels. Includes instruction in operating systems and applications; systems design and analysis; networking theory and solutions; types of networks; network management and control; network and flow optimization; security; configuring; and troubleshooting
11.9999	Computer and information sciences and support services, other	Any instructional program in computer and information sciences and support services not listed above
14.0101	Engineering, general	A program that generally prepares individuals to apply mathematical and scientific principles to solve a wide variety of practical problems in industry, social organization, public works, and commerce
14.0201	Aerospace, aeronautical and astronautical engineering	A program that prepares individuals to apply mathematical and scientific principles to the design, development and operational evaluation of aircraft, space vehicles, and their systems; applied research on flight characteristics; and the development of systems and procedures for the launching, guidance, and control of air and space vehicles
14.0801	Civil engineering, general	A program that generally prepares individuals to apply mathematical and scientific principles to the design, development and operational evaluation of structural, load-bearing, material moving, transportation, water resource, and material control systems; and environmental safety measures
14.0802	Geotechnical engineering	A program that prepares individuals to apply mathematical and scientific principles to the design, development and operational evaluation of systems for manipulating and controlling surface and subsurface features at or incorporated into structural sites, including earth and rock moving and stabilization, landfills, structural use and environmental stabilization of wastes and by-products, underground construction, and groundwater and hazardous material containment
14.0804	Transportation and highway engineering	A program that prepares individuals to apply mathematical and scientific principles to the design, development, and operational evaluation of total systems for the physical movement of people, materials and information, including general network design and planning, facilities planning, site evaluation, transportation management systems, needs projections and analysis, and analysis of costs
14.0805	Water resources engineering	A program that prepares individuals to apply mathematical and scientific principles to the design, development, and operational evaluation of systems for collecting, storing, moving, conserving and controlling surface and groundwater, including water quality control, water cycle management, management of human and industrial water requirements, water delivery, and flood control
14.0899	Civil engineering, other	Any instructional program in civil engineering not listed above
14.0901	Computer engineering, general	A program that generally prepares individuals to apply mathematical and scientific principles to the design, development, and operational evaluation of computer hardware and software systems and related equipment and facilities; and the analysis of specific problems of computer applications to various tasks
14.0902	Computer hardware engineering	(NEW) A program that prepares individuals to apply mathematical and scientific principles to the design, development, and evaluation of computer hardware and related peripheral equipment. Includes instruction in computer circuit and chip design, circuitry, computer systems design, computer equipment design, computer layout planning, testing procedures, and related computer theory and software topics
14.0903	Computer software engineering	(NEW) A program that prepares individuals to apply scientific and mathematical principles to the design, analysis, verification, validation, implementation, and maintenance of computer software systems using a variety of computer languages. Includes instruction in discrete mathematics, probability and statistics, computer science, managerial science, and applications to complex computer systems
14.0999	Computer engineering, other	(NEW) Any instructional program in computer engineering not listed above

continued

TABLE C.1 Continued

Code	Title	Description
14.1001	Electrical, electronics and communications engineering	A program that prepares individuals to apply mathematical and scientific principles to the design, development, and operational evaluation of electrical, electronic and related communications systems and their components, including electrical power generation systems; and the analysis of problems such as superconductor, wave propagation, energy storage and retrieval, and reception and amplification
14.1101	Engineering mechanics	A program with a general focus on the application of the mathematical and scientific principles of classical mechanics to the analysis and evaluation of the behavior of structures, forces and materials in engineering problems. Includes instruction in statics, kinetics, dynamics, kinematics, celestial mechanics, stress and failure, and electromagnetism
14.1201	Engineering physics	A program with a general focus on the general application of mathematical and scientific principles of physics to the analysis and evaluation of engineering problems. Includes instruction in high- and low-temperature phenomena, computational physics, superconductivity, applied thermodynamics, molecular and particle physics applications, and space science research
14.1301	Engineering science	A program with a general focus on the general application of various combinations of mathematical and scientific principles to the analysis and evaluation of engineering problems, including applied research in human behavior, statistics, biology, chemistry, the earth and planetary sciences, atmospherics and meteorology, and computer applications
14.2201	Naval architecture and marine engineering	A program that prepares individuals to apply mathematical and scientific principles to the design, development and operational evaluation of self-propelled, stationary, or towed vessels operating on or under the water, including inland, coastal, and ocean environments; and the analysis of related engineering problems such as corrosion, power transfer, pressure, hull efficiency, stress factors, safety and life support, environmental hazards and factors, and specific use requirements
14.2401	Ocean engineering	A program that prepares individuals to apply mathematical and scientific principles to the design, development, and operational evaluation of systems to monitor, control, manipulate, and operate within coastal or ocean environments, such as underwater platforms, flood control systems, dikes, hydroelectric power systems, tide and current control and warning systems, and communications equipment; the planning and design of total systems for working and functioning in water or underwater environments; and the analysis of related engineering problems such as the action of water properties and behavior on physical systems and people, tidal forces, current movements, and wave motion
14.2701	Systems engineering	A program that prepares individuals to apply mathematical and scientific principles to the design, development, and operational evaluation of total systems solutions to a wide variety of engineering problems, including the integration of human, physical, energy, communications, management, and information requirements as needed, and the application of requisite analytical methods to specific situations
14.3701	Operations research	A program that focuses on the development and application of complex mathematical or simulation models to solve problems involving operational systems, where the system concerned is subject to human intervention. Includes instruction in advanced multivariate analysis, application of judgment and statistical tests, optimization theory and techniques, resource allocation theory, mathematical modeling, control theory, statistical analysis, and applications to specific research problems
14.3801	Surveying engineering	(NEW) A program that prepares individuals to apply scientific and mathematical principles to the determination of the location, elevations, and alignment of natural and manmade topographic features. Includes instruction in property line location, surveying, surface measurement, aerial and terrestrial photogrammetry, remote sensing, satellite imagery, global positioning systems, computer applications, and photographic data processing
14.3901	Geological / geophysical engineering	(NEW) A program that prepares individuals to apply mathematical and geological principles to the analysis and evaluation of engineering problems, including the geological evaluation of construction sites, the analysis of geological forces acting on structures and systems, the analysis of potential natural resource recovery sites, and applied research on geological phenomena
14.9999	Engineering, other	Any instructional program in engineering not listed above
15.1102	Surveying technology / surveying	A program that prepares individuals to apply mathematical and scientific principles to the delineation, determination, planning, and positioning of land tracts, land and water boundaries, land contours and features; and the preparation of related maps, charts, and reports. Includes instruction in applied geodesy, computer graphics, photointerpretation, plane and geodetic surveying, mensuration, traversing, survey equipment operation and maintenance, instrument calibration, and basic cartography
15.1199	Engineering-related technologies, other	(NEW) Any programs in engineering-related technologies and technicians not listed above

continued

TABLE C.1 Continued

Code	Title	Description
15.1501	Engineering / industrial management	A program that focuses on the application of engineering principles to the planning and operational management of industrial and manufacturing operations, and prepares individuals to plan and manage such operations. Includes instruction in accounting, engineering economy, financial management, industrial and human resources management, industrial psychology, management information systems, mathematical modeling and optimization, quality control, operations research, safety and health issues, and environmental program management
16.0102	Linguistics	A program that focuses on language, language development, and relationships among languages and language groups from a humanistic and/or scientific perspective. Includes instruction in subjects such as psycholinguistics, behavioral linguistics, language acquisition, sociolinguistics, mathematical and computational linguistics, grammatical theory and theoretical linguistics, philosophical linguistics, philology and historical linguistics, comparative linguistics, phonetics, phonemics, dialectology, semantics, functional grammar and linguistics, language typology, lexicography, morphology and syntax, orthography, stylistics, structuralism, rhetoric, and applications to artificial intelligence
16.0103	Language interpretation and translation	A program that prepares individuals to be professional interpreters and/or translators of documents and data files, either from English or (Canadian) French into another language or languages or vice versa. Includes intensive instruction in one or more foreign languages plus instruction in subjects such as single- and multiple-language interpretation, one- or two-way interpretation, simultaneous interpretation, general and literary translation, business translation, technical translation, and other specific applications of linguistic skills
16.0199	Linguistic, comparative, and related language studies and services, other	(NEW) Any instructional program in linguistic, comparative, and related language studies and services not listed above
24.0101	Liberal arts and sciences / liberal studies	A program that is a structured combination of the arts, biological and physical sciences, social sciences, and humanities, emphasizing breadth of study. Includes instruction in independently designed, individualized, or regular programs
24.0102	General studies	An undifferentiated program that includes instruction in the general arts, general science, or unstructured studies
24.0103	Humanities / humanistic studies	A program that focuses on combined studies and research in the humanities subjects as distinguished from the social and physical sciences, emphasizing languages, literatures, art, music, philosophy, and religion
24.0199	Liberal arts and sciences, general studies and humanities, other	Any single instructional program in liberal arts and sciences, general studies and humanities not listed above
25.0101	Library science / librarianship	A program that focuses on the knowledge and skills required to develop, organize, store, retrieve, administer, and facilitate the use of local, remote, and networked collections of information in print, audiovisual, and electronic formats and that prepares individuals for professional service as librarians and information consultants
25.9999	Library science, other	Any instructional program in library science not listed above
27.0101	Mathematics, general	A general program that focuses on the analysis of quantities, magnitudes, forms, and their relationships, using symbolic logic and language. Includes instruction in algebra, calculus, functional analysis, geometry, number theory, logic, topology, and other mathematical specializations
27.0102	Algebra and number theory	(NEW) A program that focuses on the expression of quantities and their relationships by means of symbols, vectors, matrices, and equations, and the properties of integers. Includes instruction in algebraic structures, quadratic and automorphic forms, combinatorics, linear algebra, and algebraic geometry
27.0103	Analysis and functional analysis	(NEW) A program that focuses on the properties and behavior of equations, multivariate solutions, functions, and dynamic systems. Includes instruction in differential equations, variation, approximations, complex variables, integrals, harmonic analysis and wavelet theory, dynamic systems, and applications to mathematical physics
27.0104	Geometry / geometric analysis	(NEW) A program that focuses on the properties, measurements, and relationships pertaining to points, lines, angles, surfaces, and solids. Includes instruction in global analysis, differential geometry, Euclidian and Non-Euclidian geometry, set theory, manifolds, integral geometry, and applications to algebra and other topics
27.0105	Topology and foundations	(NEW) A program that focuses on the properties of unaltered geometric configurations under conditions of continuous, multidirectional transformations. Includes instruction in mathematical logic, proof theory, model theory, set theory, combinatorics, continua, homotopy, homology, links, and transformation actions
27.0199	Mathematics, other	(NEW) Any program in mathematics not listed above

continued

TABLE C.1 Continued

Code	Title	Description
27.0301	Applied mathematics	A program that focuses on the application of mathematics and statistics to the solution of functional problems in fields such as engineering and the applied sciences. Includes instruction in natural phenomena modeling continuum mechanics, reaction-diffusion, wave propagation, dynamic systems, numerical analysis, controlled theory, asymptotic methods, variation, optimization theory, inverse problems, and applications to specific scientific and industrial topics
27.0303	Computational mathematics	(NEW) A program that focuses on the application of mathematics to the theory, architecture, and design of computers, computational techniques, and algorithms. Includes instruction in computer theory, cybernetics, numerical analysis, algorithm development, binary structures, combinatorics, advanced statistics, and related topics
27.0399	Applied mathematics, other	Any instructional program in applied mathematics not listed above
27.0501	Statistics, general	A general program that focuses on the relationships between groups of measurements, and similarities and differences, using probability theory and techniques derived from it. Includes instruction in the principles in probability theory, binomial distribution, regression analysis, standard deviation, stochastic processes, Monte Carlo method, Bayesian statistics, nonparametric statistics, sampling theory, and statistical techniques
27.0502	Mathematical statistics and probability	(NEW) A program that focuses on the mathematical theory underlying statistical methods and their use. Includes instruction in probability theory, parametric and nonparametric inference, sequential analysis, multivariate analysis, Bayesian analysis, experimental design, time-series analysis, resampling, robust statistics, limit theory, infinite particle systems, stochastic processes, martingales, Markov processes, and Banach spaces
27.0599	Statistics, other	(NEW) Any instructional program in statistics not listed above
27.9999	Mathematics and statistics, other	Any instructional program in mathematics and statistics not listed above
29.0101	Military technologies	A program that prepares individuals to undertake advanced and specialized leadership and technical responsibilities for the armed services and related national security organizations. Includes instruction in such areas as weapons systems and technology, communications, intelligence, management, logistics, and strategy
30.0501	Peace studies and conflict resolution	A program that focuses on the origins, resolution and prevention of international and intergroup conflicts. Includes instruction in peace research methods and related social scientific and psychological knowledge bases
30.0801	Mathematics and computer science	A program with a general synthesis of mathematics and computer science or a specialization which draws from mathematics and computer science
30.2001	International / global studies	(NEW) A program that focuses on global and international issues from the perspective of the social sciences, social services, and related fields
40.0101	Physical sciences	A program that focuses on the major topics, concepts, processes, and interrelationships of physical phenomena as studied in any combination of physical science disciplines
40.0201	Astronomy	A general program that focuses on the planetary, galactic, and stellar phenomena occurring in outer space. Includes instruction in celestial mechanics, cosmology, stellar physics, galactic evolution, quasars, stellar distribution and motion, interstellar medium, atomic and molecular constituents of astronomical phenomena, planetary science, solar system evolution, and specific methodologies such as optical astronomy, radioastronomy, and theoretical astronomy
40.0202	Astrophysics	A program that focuses on the theoretical and observational study of the structure, properties, and behavior of stars, star systems and clusters, stellar life cycles, and related phenomena. Includes instruction in cosmology, plasma kinetics, stellar physics, convolution and nonequilibrium radiation transfer theory, non-Euclidean geometries, mathematical modeling, galactic structure theory, and relativistic astronomy
40.0203	Planetary astronomy and science	(NEW) A program that focuses on the scientific study of planets, small objects, and related gravitational systems. Includes instruction in the structure and composition of planetary surfaces and interiors, planetary atmospheres, satellites, orbital mechanics, asteroids and comets, solar system evolution and dynamics, planetary evolution, gravitational physics, and radiation physics
40.0401	Atmospheric sciences and meteorology, general	A general program that focuses on the scientific study of the composition and behavior of the atmospheric envelopes surrounding the Earth, the effect of Earth's atmosphere on terrestrial weather, and related problems of environment and climate. Includes instruction in atmospheric chemistry and physics, atmospheric dynamics, climatology and climate change, weather simulation, weather forecasting, climate modeling and mathematical theory; and studies of specific phenomena such as clouds, weather systems, storms, and precipitation patterns

continued

TABLE C.1 Continued

Code	Title	Description
40.0402	Atmospheric chemistry and climatology	(NEW) A program that focuses on the scientific study of atmospheric constituents, reactions, measurement techniques, and processes in predictive, current, and historical contexts. Includes instruction in climate modeling, gases and aerosols, trace gases, aqueous phase chemistry, sinks, transport mechanisms, computer measurement, climate variability, paleoclimatology, climate diagnosis, numerical modeling and data analysis, ionization, recombination, photoemission, and plasma chemistry
40.0403	Atmospheric physics and dynamics	(NEW) A program that focuses on the scientific study of the processes governing the interactions, movement, and behavior of atmospheric phenomena and related terrestrial and solar phenomena. Includes instruction in cloud and precipitation physics, solar radiation transfer, active and passive remote sensing, atmospheric electricity and acoustics, atmospheric wave phenomena, turbulence and boundary layers, solar wind, geomagnetic storms, coupling, natural plasma, and energization
40.0404	Meteorology	(NEW) A program that focuses on the scientific study of the prediction of atmospheric motion and climate change. Includes instruction in general circulation patterns, weather phenomena, atmospheric predictability, parameterization, numerical and statistical analysis, large- and mesoscale phenomena, kinematic structures, precipitation processes, and forecasting techniques
40.0601	Geology / earth science, general	A program that focuses on the scientific study of the Earth; the forces acting upon it; and the behavior of the solids, liquids, and gases comprising it. Includes instruction in historical geology, geomorphology, and sedimentology, the chemistry of rocks and soils, stratigraphy, mineralogy, petrology, geostatistics, volcanology, glaciology, geophysical principles, and applications to research and industrial problems
40.0602	Geochemistry	A program that focuses on the scientific study of the chemical properties and behavior of the silicates and other substances forming, and formed by, geomorphological processes of the Earth and other planets. Includes instruction in chemical thermodynamics, equilibrium in silicate systems, atomic bonding, isotopic fractionation, geochemical modeling, specimen analysis, and studies of specific organic and inorganic substances
40.0603	Geophysics and seismology	A program that focuses on the scientific study of the physics of solids and its application to the study of the Earth and other planets. Includes instruction in gravimetric, seismology, earthquake forecasting, magnetometry, electrical properties of solid bodies, plate tectonics, active deformation, thermodynamics, remote sensing, geodesy, and laboratory simulations of geological processes
40.0605	Hydrology and water resources science	(NEW) A program that focuses on the scientific of study of the occurrence, circulation, distribution, chemical and physical properties, and environmental interaction of surface and subsurface waters, including groundwater. Includes instruction in geophysics, thermodynamics, fluid mechanics, chemical physics, geomorphology, mathematical modeling, hydrologic analysis, continental water processes, global water balance, and environmental science
40.0606	Geochemistry and petrology	(NEW) A program that focuses on the scientific study of the igneous, metamorphic, and hydrothermal processes within the Earth and the mineral, fluid, rock, and ore deposits resulting from them. Includes instruction in mineralogy, crystallography, petrology, volcanology, economic geology, meteoritics, geochemical reactions, deposition, compound transformation, core studies, theoretical geochemistry, computer applications, and laboratory studies
40.0607	Oceanography, chemical and physical	A program that focuses on the scientific study of the chemical components, mechanisms, structure, and movement of ocean waters and their interaction with terrestrial and atmospheric phenomena. Includes instruction in material inputs and outputs, chemical and biochemical transformations in marine systems, equilibria studies, inorganic and organic ocean chemistry, oceanographic processes, sediment transport, zone processes, circulation, mixing, tidal movements, wave properties, and seawater properties
40.0699	Geological and earth sciences / geosciences, other	Any instructional program in geological and related sciences not listed above
40.0801	Physics, general	A general program that focuses on the scientific study of matter and energy, and the formulation and testing of the laws governing the behavior of the matter-energy continuum. Includes instruction in classical and modern physics, electricity and magnetism, thermodynamics, mechanics, wave properties, nuclear processes, relativity and quantum theory, quantitative methods, and laboratory methods
40.0805	Plasma and high-temperature physics	A program that focuses on the scientific study of properties and behavior of matter at high temperatures, such that molecular and atomic structures are in a disassociated ionic or electronic state. Includes instruction in magnetohydrodynamics, free electron phenomena, fusion theory, electromagnetic fields and dynamics, plasma and nonlinear wave theory, instability theory, plasma shock phenomena, quantitative modeling, and research equipment operation and maintenance

continued

TABLE C.1 Continued

Code	Title	Description
40.0807	Optics / optical sciences	A program that focuses on the scientific study of light energy, including its structure, properties, and behavior under different conditions. Includes instruction in wave theory, wave mechanics, electromagnetic theory, physical optics, geometric optics, quantum theory of light, photon detecting, laser theory, wall and beam properties, chaotic light, nonlinear optics, harmonic generation, optical systems theory, and applications to engineering problems
40.0808	Solid state and low-temperature physics	A program that focuses on the scientific study of solids and related states of matter at low energy levels, including liquids and dense gases. Includes instruction in statistical mechanics, quantum theory of solids, many-body theory, low-temperature phenomena, electron theory of metals, band theory, crystalline structures, magnetism and superconductivity, equilibria and dynamics of liquids, film and surface phenomena, quantitative modeling, and research equipment operation and maintenance
40.0809	Acoustics	A program that focuses on the scientific study of sound, and the properties and behavior of acoustic wave phenomena under different conditions. Includes instruction in wave theory, the acoustic wave equation, energy transformation, vibration phenomena, sound reflection and transmission, scattering and surface wave phenomena, singularity expansion theory, ducting, and applications to specific research problems such as underwater acoustics, crystallography, and health diagnostics
40.0810	Theoretical and mathematical physics	A program that focuses on the scientific and mathematical formulation and evaluation of the physical laws governing, and models describing, matter-energy phenomena, and the analysis of related experimental designs and results. Includes instruction in classical and quantum theory, relativity theory, field theory, mathematics of infinite series, vector and coordinate analysis, wave and particle theory, advanced applied calculus and geometry, analyses of continuum, cosmology, and statistical theory and analysis
45.0201	Anthropology	A program that focuses on the systematic study of human beings, their antecedents and related primates, and their cultural behavior and institutions, in comparative perspective. Includes instruction in biological/physical anthropology, primatology, human paleontology and prehistoric archeology, hominid evolution, anthropological linguistics, ethnography, ethnology, ethnohistory, sociocultural anthropology, psychological anthropology, research methods, and applications to areas such as medicine, forensic pathology, museum studies, and international affairs
45.0202	Physical anthropology	(NEW) A program that focuses on the application of the biological sciences and anthropology to the study of the adaptations, variability, and the evolution of human beings and their living and fossil relatives. Includes instructions in anthropology, human and mammalian anatomy, cell biology, paleontology, human culture and behavior, neuroscience, forensic anthropology, anatomical reconstruction, comparative anatomy, and laboratory science and methods
45.0299	Anthropology, other	(NEW) Any instructional program in anthropology not listed above
45.0701	Geography	A program that focuses on the systematic study of the spatial distribution and interrelationships of people, natural resources, and plant and animal life. Includes instruction in historical and political geography, cultural geography, economic and physical geography, regional science, cartographic methods, remote sensing, spatial analysis, and applications to areas such as land-use planning, development studies, and analyses of specific countries, regions, and resources
45.0702	Cartography	A program that focuses on the systematic study of map-making and the application of mathematical, computer, and other techniques to the science of mapping geographic information. Includes instruction in cartographic theory and map projections, computer-assisted cartography, map design and layout, photogrammetry, air photo interpretation, remote sensing, cartographic editing, and applications to specific industrial, commercial, research, and governmental mapping problems
45.0799	Geography, other	(NEW) Any instructional program in geography not listed above
45.0901	International relations and affairs	A program that focuses on the systematic study of international politics and institutions, and the conduct of diplomacy and foreign policy. Includes instruction in international relations theory, foreign policy analysis, national security and strategic studies, international law and organization, the comparative study of specific countries and regions, and the theory and practice of diplomacy
45.1001	Political science and government, general	A general program that focuses on the systematic study of political institutions and behavior. Includes instruction in political philosophy, political theory, comparative government and politics, political parties and interest groups, public opinion, political research methods, studies of the government and politics of specific countries, and studies of specific political institutions and processes
45.1002	American government and politics (United States)	A program that focuses on the systematic study of United States political institutions and behavior. Includes instruction in American political theory, political parties and interest groups, state and local governments, Constitutional law, federalism and national institutions, executive and legislative politics, judicial politics, popular attitudes and media influences, political research methods, and applications to the study of specific issues and institutions

continued

TABLE C.1 Continued

Code	Title	Description
45.1003	Canadian government and politics	(NEW) A program that focuses on the systematic study of Canadian political institutions and behavior. Includes instruction in British and North American political theory, political parties and interest groups, provincial and local governments, Constitutional law, federalism and national institutions, executive and legislative politics, judicial politics, popular attitudes and media influences, political research methods, and applications to the study of specific issues and institutions
45.1101	Sociology	A program that focuses on the systematic study of human social institutions and social relationships. Includes instruction in social theory, sociological research methods, social organization and structure, social stratification and hierarchies, dynamics of social change, family structures, social deviance and control, and applications to the study of specific social groups, social institutions, and social problems
45.1201	Urban studies / affairs	A program that focuses on the application of social science principles to the study of urban institutions and the forces influencing urban social and political life. Includes instruction in urban theory, the development and evolution of urban areas, urban sociology, principles of urban and social planning, and the politics and economics of urban government and services
49.0309	Marine science / Merchant Marine officer	A program that prepares individuals to serve as captains, executive officers, engineers, and ranking mates on commercially licensed inland, coastal, and ocean-going vessels. Includes instruction in maritime traditions and law; maritime policy; economics and management of commercial marine operations; basic naval architecture and engineering; shipboard power systems engineering; crew supervision; and administrative procedures
50.0401	Design and visual communications, general	A program in the applied visual arts that focuses on the general principles and techniques for effectively communicating ideas and information, and packaging products, in digital and other formats to business and consumer audiences, and that may prepare individuals in any of the applied art media
50.0409	Graphic design	(NEW) A program that prepares individuals to apply artistic and computer techniques to the interpretation of technical and commercial concepts. Includes instruction in computer-assisted art and design, printmaking, concepts sketching, technical drawing, color theory, imaging, studio technique, still and life modeling, communication skills, and commercial art business operations
50.0410	Illustration	(NEW) A program that prepares individuals to use artistic techniques to develop and execute interpretations of the concepts of authors and designers to specifications. Includes instruction in book illustration, fashion illustration, map illustration, rendering, exhibit preparation, textual layout, cartooning, and the use of various artistic techniques as requested by clients
50.0499	Design and applied arts, other	Any instructional program in design and applied arts not listed above
54.0101	History, general	A program that focuses on the general study and interpretation of the past, including the gathering, recording, synthesizing, and criticizing of evidence and theories about past events. Includes instruction in historiography; historical research methods; studies of specific periods, issues, and cultures; and applications to areas such as historic preservation, public policy, and records administration
54.0102	American history (United States)	A program that focuses on the development of American society, culture, and institutions from the Pre-Columbian period to the present. Includes instruction in American historiography, American history sources and materials, historical research methods, and applications to the study of specific themes, issues, periods, and institutions
54.0103	European history	A program that focuses on the development of European society, culture, and institutions from the origins to the present. Includes instruction in European historiography, European history sources and materials, historical research methods, and applications to the study of specific themes, issues, periods, and institutions
54.0106	Asian history	(NEW) A program that focuses on the development of the societies, cultures, and institutions of the Asian Continent from their origins to the present. Includes instruction in the historiography of specific cultures and periods; sources and materials; historical research methods; and applications to the study of specific themes, issues, periods, and institutions
54.0107	Canadian history	(NEW) A program that focuses on the study of the society, culture, and institutions of Canada from its origins to the present. Includes instruction in Canadian historiography, sources and materials, historical research methods, and applications to the study of specific themes, issues, periods, and institutions

NOTE: NEW = A new instructional program introduced in the 2000 version.
SOURCE: Department of Education Classification of Instructional Programs, 2000 version, <http://nces.ed.gov/pubs2002/cip2000/ciplist.asp>.

TABLE C.2 Instructional Programs Relevant to Core and Emerging Areas

Instructional Program	Geodesy, Geo-physics	Photo-gram-metry	Remote Sensing	Cartog-raphy	GIS, geospatial analysis	GEOINT Fusion	Crowd-sourcing	Human Geogra-phy	Visual Analytics	Fore-casting
03.0101 Natural resources / conservation, general	NOT	NOT	POS	NOT	NOT	NOT	NOT	NOT	NOT	NOT
03.0102 Environmental science / studies	NOT	NOT	POS	POS	NOT	REL	NOT	POS	NOT	REL
03.0103 Environmental studies	NOT	NOT	NOT	NOT	NOT	REL	NOT	NOT	NOT	NOT
03.0104 Environmental science	NOT	NOT	POS	POS	NOT	REL	NOT	POS	NOT	REL
03.0204 Natural resource economics	NOT	NOT	NOT	NOT	NOT	NOT	NOT	NOT	NOT	POS
03.0205 Water, wetlands, and marine resources management	NOT	NOT	POS	NOT	NOT	NOT	NOT	NOT	NOT	NOT
03.0206 Land use planning and management / development	NOT	NOT	REL	POS	POS	NOT	NOT	NOT	POS	POS
03.0501 Forestry, general	NOT	NOT	NOT	NOT	NOT	REL	NOT	NOT	NOT	REL
03.0502 Forest sciences and biology	NOT	NOT	POS	NOT	NOT	NOT	NOT	NOT	NOT	NOT
03.0506 Forest management / forest resources management	NOT	POS	REL	POS	REL	REL	NOT	NOT	NOT	NOT
03.0508 Urban forestry	NOT	NOT	POS	NOT	NOT	NOT	NOT	NOT	NOT	NOT
03.0509 Wood science and wood products / pulp and paper technology	NOT	NOT	POS	NOT	NOT	NOT	NOT	NOT	NOT	NOT
04.0301 City / urban, community and regional planning	NOT	NOT	REL	POS	REL	NOT	NOT	NOT	POS	POS
05.0101 African studies	NOT	NOT	NOT	POS	NOT	NOT	NOT	REL	NOT	NOT
05.0102 American / United States studies / civilization	NOT	NOT	NOT	NOT	NOT	NOT	NOT	REL	NOT	NOT
05.0103 Asian studies / civilization	NOT	NOT	NOT	POS	NOT	NOT	NOT	REL	NOT	NOT
05.0104 East Asian studies	NOT	NOT	NOT	POS	NOT	NOT	NOT	REL	NOT	NOT
05.0105 Central / middle and eastern European studies	NOT	NOT	NOT	POS	NOT	NOT	NOT	REL	NOT	NOT
05.0106 European studies / civilization	NOT	NOT	NOT	NOT	NOT	NOT	NOT	REL	NOT	NOT
05.0107 Latin American studies	NOT	NOT	NOT	POS	NOT	NOT	NOT	REL	NOT	NOT
05.0108 Near and Middle Eastern studies	NOT	NOT	NOT	POS	NOT	NOT	NOT	REL	NOT	NOT
05.0109 Pacific area / Pacific Rim studies	NOT	NOT	NOT	POS	NOT	NOT	NOT	REL	NOT	NOT
05.0110 Russian studies	NOT	NOT	NOT	POS	NOT	NOT	NOT	REL	NOT	NOT
05.0111 Scandinavian studies	NOT	NOT	NOT	NOT	NOT	NOT	NOT	REL	NOT	NOT
05.0112 South Asian studies	NOT	NOT	NOT	POS	NOT	NOT	NOT	REL	NOT	NOT
05.0113 Southeast Asian studies	NOT	NOT	NOT	POS	NOT	NOT	NOT	REL	NOT	NOT
05.0114 Western European studies	NOT	NOT	NOT	NOT	NOT	NOT	NOT	REL	NOT	NOT

continued

TABLE C.2 Continued

Instructional Program	Geodesy, Geo-physics	Photo-gram-metry	Remote Sensing	Cartog-raphy	GIS, geospatial analysis	GEOINT Fusion	Crowd-sourcing	Human Geogra-phy	Visual Analytics	Fore-casting
05.0115 Canadian studies	NOT	NOT	NOT	NOT	NOT	NOT	NOT	REL	NOT	NOT
05.0116 Balkans studies	NOT	NOT	NOT	POS	NOT	NOT	NOT	REL	NOT	NOT
05.0117 Baltic studies	NOT	NOT	NOT	POS	NOT	NOT	NOT	REL	NOT	NOT
05.0118 Slavic studies	NOT	NOT	NOT	POS	NOT	NOT	NOT	REL	NOT	NOT
05.0119 Caribbean studies	NOT	NOT	NOT	POS	NOT	NOT	NOT	REL	NOT	NOT
05.0120 Ural-Altaic and Central Asian studies	NOT	NOT	NOT	POS	NOT	NOT	NOT	REL	NOT	NOT
05.0121 Commonwealth studies	NOT	NOT	NOT	NOT	NOT	NOT	NOT	REL	NOT	NOT
05.0122 Regional studies (U.S., Canadian, foreign)	NOT	NOT	NOT	NOT	NOT	NOT	NOT	REL	NOT	NOT
05.0123 Chinese studies	NOT	NOT	NOT	POS	NOT	NOT	NOT	REL	NOT	NOT
05.0124 French studies	NOT	NOT	NOT	NOT	NOT	NOT	NOT	REL	NOT	NOT
05.0125 German studies	NOT	NOT	NOT	NOT	NOT	NOT	NOT	REL	NOT	NOT
05.0126 Italian studies	NOT	NOT	NOT	NOT	NOT	NOT	NOT	REL	NOT	NOT
05.0127 Japanese studies	NOT	NOT	NOT	NOT	NOT	NOT	NOT	REL	NOT	NOT
05.0128 Korean studies	NOT	NOT	NOT	POS	NOT	NOT	NOT	REL	NOT	NOT
05.0129 Polish studies	NOT	NOT	NOT	NOT	NOT	NOT	NOT	REL	NOT	NOT
05.0130 Spanish and Iberian studies	NOT	NOT	NOT	NOT	NOT	NOT	NOT	REL	NOT	NOT
05.0131 Tibetan studies	NOT	NOT	NOT	POS	NOT	NOT	NOT	REL	NOT	NOT
05.0132 Ukraine studies	NOT	NOT	NOT	POS	NOT	NOT	NOT	REL	NOT	NOT
05.0199 Area studies, other	NOT	NOT	NOT	NOT	NOT	NOT	NOT	REL	NOT	NOT
10.0301 Graphic communications, general	NOT	NOT	NOT	POS	POS	NOT	NOT	NOT	REL	NOT
10.0302 Printing management	NOT	NOT	NOT	POS	NOT	NOT	NOT	NOT	NOT	NOT
10.0303 Prepress / desktop publishing and digital imaging design	NOT	NOT	POS	REL	NOT	NOT	NOT	POS	REL	NOT
10.0304 Animation, interactive technology, video graphics and special effects	NOT	NOT	POS	REL	POS	NOT	NOT	NOT	REL	POS
10.0308 Computer typography and composition equipment operator	NOT	NOT	NOT	POS	NOT	NOT	NOT	NOT	POS	NOT
10.0399 Graphic communications, other	NOT	NOT	NOT	POS	NOT	NOT	NOT	NOT	REL	NOT
11.0101 Computer and information sciences, general	NOT	NOT	NOT	POS	NOT	NOT	POS	POS	POS	NOT
11.0102 Artificial intelligence and robotics	NOT	NOT	POS	NOT	NOT	REL	POS	POS	NOT	REL
11.0103 Information technology	NOT	NOT	POS	POS	POS	REL	REL	POS	POS	NOT
11.0199 Computer and information sciences, other	NOT	NOT	POS	NOT	NOT	NOT	REL	POS	POS	NOT
11.0201 Computer programming / programmer, general	NOT	NOT	NOT	REL	NOT	NOT	POS	POS	POS	POS

continued

TABLE C.2 Continued

Instructional Program	Geodesy, Geo-physics	Photo-gram-metry	Remote Sensing	Cartog-raphy	GIS, geospatial analysis	GEOINT Fusion	Crowd-sourcing	Human Geogra-phy	Visual Analytics	Fore-casting
11.0202 Computer programming, specific applications	POS	POS	POS	REL	POS	NOT	NOT	NOT	REL	POS
11.0203 Computer programming, vendor / product certification	NOT	NOT	REL	REL	NOT	NOT	NOT	NOT	POS	POS
11.0299 Computer programming, other	NOT	NOT	NOT	NOT	NOT	NOT	POS	NOT	POS	NOT
11.0301 Data processing and data processing technology / technician	NOT	NOT	POS	POS	NOT	NOT	NOT	POS	POS	NOT
11.0401 Information science / studies	NOT	NOT	POS	POS	POS	REL	POS	REL	REL	POS
11.0501 Computer systems analysis / analyst	NOT	NOT	POS	NOT	NOT	NOT	POS	NOT	POS	POS
11.0701 Computer science	NOT	NOT	POS	NOT	POS	NOT	POS	NOT	POS	POS
11.0801 Web page, digital / multimedia and information resources design	NOT	NOT	NOT	POS	POS	NOT	POS	REL	REL	NOT
11.0802 Data modeling / warehousing and database administration	NOT	NOT	NOT	NOT	POS	REL	POS	POS	POS	NOT
11.0803 Computer graphics	NOT	NOT	POS	POS	POS	NOT	POS	NOT	REL	NOT
11.0899 Computer software and media applications, other	NOT	NOT	NOT	POS	NOT	NOT	POS	NOT	POS	NOT
11.0901 Computer systems networking and telecommunications	NOT	NOT	POS	NOT	NOT	NOT	POS	NOT	POS	NOT
11.9999 Computer and information sciences and support services, other	NOT	NOT	POS	NOT	NOT	NOT	POS	NOT	NOT	POS
14.0101 Engineering, general	NOT	NOT	NOT	NOT	NOT	NOT	POS	NOT	NOT	NOT
14.0201 Aerospace, aeronautical and astronautical engineering	REL	NOT	POS	NOT	NOT	NOT	POS	NOT	NOT	NOT
14.0801 Civil engineering, general	NOT	NOT	NOT	NOT	NOT	NOT	POS	NOT	NOT	NOT
14.0802 Geotechnical engineering	NOT	NOT	NOT	NOT	NOT	NOT	POS	NOT	NOT	NOT
14.0804 Transportation and highway engineering	NOT	NOT	POS	NOT	POS	NOT	POS	NOT	NOT	POS
14.0805 Water resources engineering	NOT	NOT	POS	NOT	NOT	NOT	POS	NOT	NOT	NOT
14.0899 Civil engineering, other	NOT	POS	NOT	NOT	NOT	NOT	POS	NOT	NOT	NOT
14.0901 Computer engineering, general	NOT	NOT	NOT	NOT	NOT	NOT	POS	NOT	POS	POS
14.0902 Computer hardware engineering	NOT	NOT	NOT	NOT	NOT	NOT	POS	NOT	NOT	NOT

continued

TABLE C.2 Continued

Instructional Program	Geodesy, Geophysics	Photogrammetry	Remote Sensing	Cartography	GIS, geospatial analysis	GEOINT Fusion	Crowdsourcing	Human Geography	Visual Analytics	Forecasting
14.0903 Computer software engineering	NOT	NOT	POS	POS	POS	NOT	POS	POS	POS	REL
14.0999 Computer engineering, other	NOT	NOT	NOT	NOT	NOT	NOT	POS	NOT	POS	NOT
14.1001 Electrical, electronics and communications engineering	NOT	NOT	NOT	NOT	NOT	NOT	POS	NOT	NOT	NOT
14.1101 Engineering mechanics	POS	NOT	NOT	NOT	NOT	NOT	NOT	NOT	NOT	REL
14.1201 Engineering physics	REL	POS	POS	NOT	NOT	NOT	NOT	NOT	NOT	POS
14.1301 Engineering science	REL	POS	NOT	NOT	NOT	NOT	POS	NOT	REL	REL
14.2201 Naval architecture and marine engineering	NOT	NOT	POS	NOT	NOT	NOT	NOT	NOT	NOT	NOT
14.2401 Ocean engineering	NOT	NOT	NOT	NOT	NOT	NOT	NOT	NOT	NOT	POS
14.2701 Systems engineering	NOT	NOT	NOT	NOT	NOT	NOT	NOT	NOT	POS	NOT
14.3701 Operations research	NOT	NOT	NOT	NOT	REL	NOT	POS	POS	NOT	REL
14.3801 Surveying engineering	REL	REL	REL	REL	NOT	REL	NOT	NOT	POS	NOT
14.3901 Geological / geophysical engineering	NOT	NOT	REL	NOT	NOT	NOT	POS	NOT	NOT	POS
14.9999 Engineering, other	NOT	NOT	NOT	NOT	NOT	NOT	NOT	NOT	POS	NOT
15.1102 Surveying technology / surveying	POS	POS	REL	REL	POS	REL	POS	NOT	POS	NOT
15.1199 Engineering-related technologies, other	NOT	NOT	NOT	NOT	NOT	NOT	NOT	NOT	POS	NOT
15.1501 Engineering / industrial management	NOT	NOT	NOT	NOT	NOT	NOT	NOT	NOT	NOT	REL
16.0102 Linguistics	NOT	NOT	NOT	NOT	NOT	NOT	POS	NOT	POS	NOT
16.0103 Language interpretation and translation	NOT	NOT	NOT	POS	NOT	NOT	POS	REL	NOT	NOT
16.0199 Linguistic, comparative, and related language studies and services, other	NOT	NOT	NOT	POS	NOT	NOT	POS	POS	NOT	NOT
24.0101 Liberal arts and sciences / liberal studies	NOT	NOT	NOT	NOT	NOT	NOT	NOT	POS	NOT	NOT
24.0102 General studies	NOT	NOT	NOT	NOT	NOT	NOT	NOT	POS	NOT	NOT
24.0103 Humanities / humanistic studies	NOT	NOT	NOT	NOT	NOT	NOT	NOT	REL	NOT	NOT
24.0199 Liberal arts and sciences, general studies and humanities, other	NOT	NOT	NOT	NOT	NOT	NOT	NOT	POS	NOT	NOT
25.0101 Library science / librarianship	NOT	NOT	NOT	NOT	NOT	NOT	NOT	REL	NOT	NOT

continued

TABLE C.2 Continued

Instructional Program	Geodesy, Geo-physics	Photo-gram-metry	Remote Sensing	Cartog-raphy	GIS, geospatial analysis	GEOINT Fusion	Crowd-sourcing	Human Geogra-phy	Visual Analytics	Fore-casting
25.9999 Library science, other	NOT	NOT	NOT	POS	NOT	NOT	NOT	POS	NOT	NOT
27.0101 Mathematics, general	NOT	NOT	REL	POS	NOT	NOT	NOT	NOT	POS	POS
27.0102 Algebra and number theory	NOT	POS	POS	NOT	NOT	NOT	NOT	NOT	NOT	REL
27.0103 Analysis and functional analysis	POS	NOT	POS	POS	NOT	NOT	POS	NOT	NOT	NOT
27.0104 Geometry / geometric analysis	NOT	POS	POS	REL	POS	NOT	NOT	NOT	POS	NOT
27.0105 Topology and foundations	NOT	NOT	REL	POS	NOT	NOT	NOT	NOT	NOT	NOT
27.0199 Mathematics, other	NOT	NOT	NOT	NOT	NOT	NOT	POS	NOT	NOT	NOT
27.0301 Applied mathematics	POS	NOT	NOT	POS	NOT	NOT	POS	REL	POS	REL
27.0303 Computational mathematics	NOT	NOT	POS	POS	POS	NOT	POS	REL	POS	REL
27.0399 Applied mathematics, other	NOT	NOT	NOT	POS	NOT	NOT	POS	POS	POS	NOT
27.0501 Statistics, general	NOT	POS	REL	POS	POS	NOT	REL	REL	POS	REL
27.0502 Mathematical statistics and probability	POS	POS	REL	NOT	POS	REL	REL	REL	POS	REL
27.0599 Statistics, other	NOT	NOT	NOT	POS	NOT	NOT	REL	REL	POS	NOT
27.9999 Mathematics and statistics, other	NOT	NOT	NOT	POS	NOT	NOT	POS	POS	NOT	NOT
29.0101 Military technologies	NOT	NOT	REL	POS	NOT	NOT	NOT	POS	NOT	NOT
30.0501 Peace studies and conflict resolution	NOT	NOT	NOT	NOT	NOT	NOT	NOT	POS	NOT	NOT
30.0801 Mathematics and computer science	NOT	NOT	REL	NOT	NOT	NOT	NOT	POS	POS	REL
30.2001 International / global studies	NOT	NOT	POS	NOT	NOT	NOT	NOT	REL	NOT	NOT
40.0101 Physical sciences	POS	NOT	REL	NOT	NOT	NOT	NOT	NOT	NOT	NOT
40.0201 Astronomy	POS	NOT	NOT	NOT	NOT	NOT	NOT	NOT	NOT	NOT
40.0202 Astrophysics	NOT	NOT	NOT	NOT	NOT	NOT	NOT	NOT	NOT	REL
40.0203 Planetary astronomy and science	POS	NOT	NOT	NOT	NOT	NOT	NOT	NOT	NOT	NOT
40.0401 Atmospheric sciences and meteorology, general	POS	NOT	POS	NOT	NOT	REL	NOT	NOT	NOT	REL
40.0402 Atmospheric chemistry and climatology	NOT	NOT	NOT	NOT	NOT	NOT	NOT	NOT	NOT	REL
40.0403 Atmospheric physics and dynamics	POS	NOT	REL	NOT	NOT	NOT	NOT	NOT	NOT	REL
40.0404 Meteorology	NOT	NOT	POS	NOT	NOT	NOT	NOT	NOT	NOT	REL
40.0601 Geology / earth science, general	POS	NOT	REL	NOT	NOT	NOT	NOT	NOT	NOT	REL
40.0602 Geochemistry	NOT	NOT	POS	NOT	NOT	NOT	NOT	NOT	NOT	NOT
40.0603 Geophysics and seismology	REL	NOT	REL	NOT	NOT	NOT	NOT	NOT	NOT	REL
40.0605 Hydrology and water resources science	NOT	NOT	POS	NOT	NOT	NOT	NOT	NOT	NOT	REL
40.0606 Geochemistry and petrology	NOT	NOT	POS	NOT	NOT	NOT	NOT	NOT	NOT	NOT

continued

TABLE C.2 Continued

Instructional Program	Geodesy, Geophysics	Photogrammetry	Remote Sensing	Cartography	GIS, geospatial analysis	GEOINT Fusion	Crowd-sourcing	Human Geography	Visual Analytics	Forecasting
40.0607 Oceanography, chemical and physical	NOT	NOT	POS	NOT	NOT	NOT	NOT	NOT	NOT	REL
40.0699 Geological and earth sciences / geosciences, other	NOT	NOT	NOT	NOT	NOT	NOT	NOT	POS	NOT	NOT
40.0801 Physics, general	POS	NOT	POS	NOT	NOT	NOT	NOT	NOT	NOT	REL
40.0805 Plasma and high-temperature physics	NOT	NOT	NOT	NOT	NOT	NOT	NOT	NOT	NOT	REL
40.0807 Optics / optical sciences	POS	POS	REL	NOT	NOT	NOT	NOT	NOT	NOT	NOT
40.0808 Solid state and low-temperature physics	NOT	NOT	POS	NOT	NOT	NOT	NOT	NOT	NOT	REL
40.0809 Acoustics	NOT	NOT	POS	NOT	NOT	NOT	POS	NOT	NOT	NOT
40.0810 Theoretical and mathematical physics	POS	NOT	NOT	NOT	NOT	NOT	POS	NOT	NOT	REL
45.0201 Anthropology	NOT	NOT	POS	NOT	NOT	NOT	NOT	REL	NOT	NOT
45.0202 Physical anthropology	NOT	NOT	POS	NOT	NOT	NOT	NOT	REL	NOT	NOT
45.0299 Anthropology, other	NOT	NOT	NOT	NOT	NOT	NOT	NOT	REL	NOT	NOT
45.0701 Geography	NOT	NOT	REL	REL	REL	NOT	POS	POS	POS	POS
45.0702 Cartography	NOT	POS	REL	REL	REL	REL	POS	NOT	REL	NOT
45.0799 Geography, other	NOT	NOT	REL	POS	NOT	NOT	NOT	NOT	POS	NOT
45.0901 International relations and affairs	NOT	NOT	NOT	NOT	NOT	NOT	NOT	POS	NOT	POS
45.1001 Political science and government, general	NOT	NOT	NOT	NOT	NOT	NOT	NOT	REL	NOT	REL
45.1002 American government and politics (United States)	NOT	NOT	NOT	NOT	NOT	NOT	NOT	POS	NOT	NOT
45.1003 Canadian government and politics	NOT	NOT	NOT	NOT	NOT	NOT	NOT	POS	NOT	NOT
45.1101 Sociology	NOT	NOT	NOT	NOT	NOT	NOT	NOT	REL	NOT	REL
45.1201 Urban studies / affairs	NOT	NOT	POS	NOT	POS	NOT	NOT	POS	NOT	NOT
49.0309 Marine science / Merchant Marine officer	NOT	NOT	POS	NOT	NOT	NOT	NOT	NOT	NOT	NOT
50.0401 Design and visual communications, general	NOT	NOT	NOT	REL	POS	NOT	NOT	NOT	POS	NOT
50.0409 Graphic design	NOT	NOT	NOT	REL	POS	NOT	NOT	NOT	REL	NOT
50.0410 Illustration	NOT	NOT	NOT	NOT	NOT	NOT	NOT	NOT	POS	NOT
50.0499 Design and applied arts, other	NOT	NOT	NOT	NOT	NOT	NOT	NOT	NOT	POS	NOT
54.0101 History, general	NOT	NOT	NOT	NOT	NOT	NOT	NOT	REL	NOT	NOT
54.0102 American history (United States)	NOT	NOT	NOT	NOT	NOT	NOT	NOT	REL	NOT	NOT
54.0103 European history	NOT	NOT	NOT	NOT	NOT	NOT	NOT	REL	NOT	NOT
54.0106 Asian history	NOT	NOT	NOT	NOT	NOT	NOT	NOT	REL	NOT	NOT
54.0107 Canadian history	NOT	NOT	NOT	NOT	NOT	NOT	NOT	REL	NOT	NOT

NOTE: "REL" indicates that the field is highly relevant to the area; "POS" indicates that the field is possibly relevant to the area; "NOT" indicates that the field is not relevant to the area.

SOURCE: Programs are based on the 2000 version of the Classification of Instructional Programs developed by the National Center for Education Statistics (<http://nces.ed.gov/pubs2002/cip2000/ciplist.asp>).

TABLE C.3 Number of Degrees Conferred in 2009 Across Fields of Study That Are Relevant to the Core and Emerging Areas

| Area | Bachelor's Degrees | | Master's Degrees | | Doctorate Degrees | | Total Degrees | |
	Highly Relevant	Possibly Relevant	Highly Relevant	Possibly Relevant	Highly Relevant	Possibly Relevant	Highly Relevant	Possibly Relevant
Geodesy and geophysics	3,963	10,333	1,578	3,742	438	2,322	5,979	16,397
Photogrammetry	21	1,946	6	2,162	1	529	28	4,637
Remote sensing	24,051	47,484	9,231	18,866	2,145	3,764	35,427	70,114
Cartography	12,986	43,761	1,566	20,367	227	2,600	14,779	66,728
GIS and geospatial analysis	5,971	26,941	3,574	13,719	372	1,312	9,917	41,972
GEOINT Fusion	14,806		6,296		554		21,656	
Crowdsourcing	3,427	70,088	2,658	38,100	384	5,859	6,469	114,047
Human geography	128,483	78,090	22,856	18,284	3,677	1,333	155,016	97,707
Visual analytics	13,053	57,577	4,414	27,370	211	3,946	17,678	88,893
Forecasting	83,031	47,719	13,614	21,460	4,476	2,552	101,121	71,731

SOURCE: Data from the U.S. Department of Education, National Center for Education Statistics' Integrated Postsecondary Education Data System Completions Survey. Accessed via WebCASPAR.

TABLE C.4 Number of Degrees Conferred by Year Across Fields of Study That Are Highly Relevant to the Core and Emerging Areas

Degree Level	2000	2001	2002	2003	2004	2005	2006	2007	2008	2009
Bachelor's degrees	132,421	132,907	138,943	153,241	164,028	170,740	175,662	178,861	179,287	181,062
Master's degrees	27,355	27,756	28,981	31,516	34,766	35,643	36,870	36,527	39,077	40,737
Doctorate degrees	6,603	6,361	6,196	6,316	6,418	6,569	7,012	7,277	7,514	7,989
TOTAL	166,379	167,024	174,120	191,073	205,212	212,952	219,544	222,665	225,878	229,788

TABLE C.5 Number of Degrees Conferred by Year and Area Across Highly Relevant Fields of Study

Area	Degree	2000	2001	2002	2003	2004	2005	2006	2007	2008	2009
Geodesy and geophysics	Bachelor's	1,851	2,045	2,362	2,586	2,984	3,089	3,482	3,660	3,793	3,963
	Master's	949	962	1,119	1,236	1,327	1,449	1,549	1,374	1,548	1,578
	Doctorate	318	338	318	297	339	352	391	437	405	438
	Total	3,118	3,345	3,799	4,119	4,650	4,890	5,422	5,471	5,746	5,979
Photogrammetry	Bachelor's	—	—	—	—	7	23	21	34	36	21
	Master's	—	—	—	—	8	20	4	5	8	6
	Doctorate	—	—	—	—	2	3	1	3		1
	Total	—	—	—	—	17	46	26	42	44	28
Remote sensing	Bachelor's	19,484	19,083	19,676	20,179	21,071	21,866	22,249	22,948	23,363	24,051
	Master's	6,286	5,988	6,264	6,836	7,524	8,460	8,758	8,515	8,975	9,231
	Doctorate	1,637	1,522	1,509	1,513	1,601	1,703	1,840	1,987	1,980	2,145
	Total	27,407	26,593	27,449	28,528	30,196	32,029	32,847	33,450	34,318	35,427
Cartographic science	Bachelor's	5,940	6,107	6,519	9,105	11,404	11,052	11,607	12,194	12,671	12,986
	Master's	1,016	1,001	1,026	1,350	1,356	1,459	1,473	1,428	1,602	1,566
	Doctorate	204	212	219	195	217	222	230	221	268	227
	Total	7,160	7,320	7,764	10,650	12,977	12,733	13,310	13,843	14,541	14,779
GIS and geospatial analysis	Bachelor's	5,427	5,297	5,096	5,079	5,449	5,409	5,372	5,632	5,605	5,971
	Master's	2,380	2,220	2,338	2,743	2,915	3,314	3,293	3,226	3,401	3,574
	Doctorate	305	298	339	304	343	327	366	345	396	372
	Total	8,112	7,815	7,773	8,126	8,707	9,050	9,031	9,203	9,402	9,917
GEOINT fusion	Bachelor's	13,098	14,208	15,620	18,225	19,161	18,300	17,928	16,912	14,259	14,806
	Master's	3,805	4,291	4,637	5,173	5,839	4,957	5,047	4,828	5,814	6,296
	Doctorate	297	293	307	369	379	398	455	446	527	554
	Total	17,200	18,792	20,564	23,767	25,379	23,655	23,430	22,186	20,600	21,656
Crowdsourcing	Bachelor's	380	371	388	1,487	5,548	5,744	5,894	5,824	3,161	3,427
	Master's	783	805	919	1,336	2,414	1,853	1,905	2,092	2,412	2,658
	Doctorate	220	201	178	209	220	275	278	316	310	384
	Total	1,383	1,377	1,485	3,032	8,182	7,872	8,077	8,232	5,883	6,469
Human geography	Bachelor's	99,924	100,815	104,998	114,063	116,872	122,805	125,650	127,047	128,654	128,483
	Master's	16,782	17,196	17,976	18,862	20,167	19,995	20,825	20,928	22,085	22,856
	Doctorate	3,322	3,185	3,142	3,187	3,208	3,112	3,294	3,265	3,401	3,677
	Total	120,028	121,196	126,116	136,112	140,247	145,912	149,769	151,240	154,140	155,016
Visual analytics	Bachelor's	7,535	8,933	10,222	13,766	12,398	11,983	11,807	11,717	12,639	13,053
	Master's	2,656	3,158	3,593	4,123	3,769	3,311	3,513	3,197	3,855	4,414
	Doctorate	105	91	105	107	157	137	181	177	211	211
	Total	10,296	12,182	13,920	17,996	16,324	15,431	15,501	15,091	16,705	17,678
Forecasting	Bachelor's	67,884	67,319	69,059	73,070	75,327	79,455	80,861	82,292	83,071	83,031
	Master's	9,972	10,014	10,510	11,065	11,920	12,273	12,319	12,320	12,852	13,614
	Doctorate	3,615	3,554	3,328	3,509	3,427	3,650	3,887	4,078	4,160	4,476
	Total	81,471	80,887	82,897	87,644	90,674	95,378	97,067	98,690	100,083	101,121

SOURCE: Data from the U.S. Department of Education, National Center for Education Statistics' Integrated Postsecondary Education Data System Completions Survey. Accessed via WebCASPAR.

TABLE C.6 Number of Degrees Conferred by Year in Each Field of Study That Is Highly Relevant to a Core or Emerging Area

Instructional Program	2000	2001	2002	2003	2004	2005	2006	2007	2008	2009
Geodesy and geophysics										
14.0201 Aerospace, aeronautical and astronautical engineering	2,030	2,311	2,622	2,932	3,378	3,535	4,058	4,100	4,308	4,476
14.1201 Engineering physics	322	306	325	349	408	443	470	526	534	510
14.1301 Engineering science	599	537	655	656	644	663	674	602	643	752
14.3801 Surveying engineering	—	—	—	—	17	46	26	42	44	28
40.0603 Geophysics and seismology	167	191	197	182	203	203	194	201	217	213
Total	3,118	3,345	3,799	4,119	4,650	4,890	5,422	5,471	5,746	5,979
Photogrammetry										
14.3801 Surveying engineering	—	—	—	—	17	46	26	42	44	28
Total	—	—	—	—	17	46	26	42	44	28
Remote sensing										
03.0206 Land use planning and management / development	—	—	—	29	100	84	90	109	96	132
03.0506 Forest management / forest resources management	262	211	201	188	164	150	128	132	148	149
04.0301 City / urban, community and regional planning	1,914	1,815	1,975	2,221	2,413	2,653	2,804	2,859	2,839	3,047
11.0203 Computer programming, vendor / product certification	—	—	—	—	—	—	—	—	—	—
14.3801 Surveying engineering	—	—	—	—	17	46	26	42	44	28
14.3901 Geological / geophysical engineering	239	192	178	168	165	139	132	173	186	215
15.1102 Survey technology / surveying	207	247	210	217	202	207	206	242	257	259
27.0101 Mathematics, general	12,863	12,383	13,097	13,562	14,536	15,654	16,416	16,745	16,799	17,021
27.0105 Topology and foundations	—	—	—	—	—	—	—	—	—	—
27.0501 Statistics, general	1,383	1,377	1,485	1,656	1,931	2,024	2,009	2,135	2,248	2,392
27.0502 Mathematical statistics and probability	—	—	—	6	4	19	16	15	100	110
29.0101 Military technologies	154	21	3	48	10	316	363	59	408	482
30.0801 Mathematics and computer science	375	371	471	456	437	366	250	230	188	199
40.0101 Physical sciences	392	365	349	332	268	353	315	297	341	357
40.0403 Atmospheric physics and dynamics	—	—	—	—	6	3	9	8	7	7
40.0601 Geology / earth science, general	4,358	4,289	4,243	4,219	4,144	4,124	4,290	4,358	4,431	4,681
40.0603 Geophysics and seismology	167	191	197	182	203	203	194	201	217	213
40.0807 Optics / optical sciences	59	63	62	128	148	159	167	171	175	171
45.0701 Geography	4,930	4,928	4,885	4,965	5,245	5,273	5,196	5,364	5,480	5,615
45.0702 Cartography	104	140	93	99	124	135	117	135	126	165
45.0799 Geography, other	—	—	—	52	79	121	119	175	228	184
Total	27,407	26,593	27,449	28,528	30,196	32,029	32,847	33,450	34,318	35,427

continued

TABLE C.6 Continued

Instructional Program	2000	2001	2002	2003	2004	2005	2006	2007	2008	2009
Cartographic science										
10.0303 Prepress / desktop publishing and digital imaging design	—	—	—	—	23	34	28	54	46	71
10.0304 Animation, interactive technology, video graphics and special effects	—	—	—	169	379	563	873	1,232	1,628	1,675
11.0201 Computer programming / programmer, general	525	539	772	914	748	542	452	448	407	365
11.0202 Computer programming special applications	—	—	—	405	212	66	77	68	44	43
11.0203 Computer programming, vendor / product certification	—	—	—	—	—	—	—	—	—	—
14.3801 Surveying engineering	—	—	—	—	17	46	26	42	44	28
15.1102 Survey technology / surveying	207	247	210	217	202	207	206	242	257	259
27.0104 Geometry / geometric analysis	—	—	—	—	—	—	—	—	—	—
45.0701 Geography	4,930	4,928	4,885	4,965	5,245	5,273	5,196	5,364	5,480	5,615
45.0702 Cartography	104	140	93	99	124	135	117	135	126	165
50.0401 Design and visual communications, general	1,394	1,466	1,804	2,424	3,072	2,513	2,715	2,524	2,439	2,213
50.0409 Graphic design	—	—	—	1,457	2,955	3,354	3,620	3,734	4,070	4,345
Total	7,160	7,320	7,764	10,650	12,977	12,733	13,310	13,843	14,541	14,779
GIS and geospatial analysis										
03.0506 Forest management / forest resources management	262	211	201	188	164	150	128	132	148	149
04.0301 City / urban, community and regional planning	1,914	1,815	1,975	2,221	2,413	2,653	2,804	2,859	2,839	3,047
14.3701 Operations research	902	721	619	653	761	839	786	713	809	941
45.0701 Geography	4,930	4,928	4,885	4,965	5,245	5,273	5,196	5,364	5,480	5,615
45.0702 Cartography	104	140	93	99	124	135	117	135	126	165
Total	8,112	7,815	7,773	8,126	8,707	9,050	9,031	9,203	9,402	9,917
GEOINT fusion										
03.0102 Environmental science / studies	5,186	4,942	5,180	3,029	391	—	—	—	—	—
03.0103 Environmental studies	—	—	—	1,325	3,070	3,261	3,367	3,292	3,571	3,562
03.0104 Environmental science	—	—	—	1,123	2,410	2,561	2,557	2,680	2,800	3,214
03.0501 Forestry, general	1,115	1,081	974	991	849	1,006	1,084	930	939	971
03.0506 Forest management / forest resources management	262	211	201	188	164	150	128	132	148	149
11.0102 Artificial intelligence and robotics	—	—	—	16	45	47	72	50	97	84
11.0103 Information technology	—	—	—	1,194	6,082	5,681	5,907	5,932	3,395	3,764
11.0401 Information science / studies	9,593	11,505	13,172	14,600	11,149	9,655	9,108	7,856	8,295	8,452
11.0802 Data modeling / warehousing and database administration	—	—	—	212	148	46	77	68	45	68
14.3801 Surveying engineering	—	—	—	—	17	46	26	42	44	28
15.1102 Survey technology / surveying	207	247	210	217	202	207	206	242	257	259
27.0502 Mathematical statistics and probability	—	—	—	6	4	19	16	15	100	110
40.0401 Atmospheric sciences and meteorology, general	733	666	734	767	724	841	765	812	783	830
45.0702 Cartography	104	140	93	99	124	135	117	135	126	165
Total	17,200	18,792	20,564	23,767	25,379	23,655	23,430	22,186	20,600	21,656

continued

TABLE C.6 Continued

Instructional Program	2000	2001	2002	2003	2004	2005	2006	2007	2008	2009
Crowdsourcing										
11.0103 Information technology	—	—	—	1,194	6,082	5,681	5,907	5,932	3,395	3,764
11.0199 Computer and information science, other	—	—	—	160	155	136	131	120	116	165
27.0501 Statistics, general	1,383	1,377	1,485	1,656	1,931	2,024	2,009	2,135	2,248	2,392
27.0502 Mathematical statistics and probability	—	—	—	6	4	19	16	15	100	110
27.0599 Statistics, other	—	—	—	16	10	12	14	30	24	38
Total	1,383	1,377	1,485	3,032	8,182	7,872	8,077	8,232	5,883	6,469
Human geography										
05.0101 African studies	69	42	53	74	80	68	87	61	112	135
05.0102 American / United States studies / civilization	1,813	1,918	1,934	1,997	1,971	2,025	2,018	1,978	1,987	1,932
05.0103 Asian studies / civilization	586	513	529	522	556	627	701	738	810	857
05.0104 East Asian studies	508	402	437	418	453	441	526	541	589	621
05.0105 Central / middle and eastern European studies	8	30	42	46	15	21	22	16	34	34
05.0106 European studies / civilization	106	133	140	127	116	140	198	178	176	193
05.0107 Latin American studies	694	633	618	731	770	801	806	834	825	822
05.0108 Near and Middle Eastern studies	221	184	176	235	301	317	332	305	353	412
05.0109 Pacific area / Pacific Rim studies	12	13	16	11	17	26	24	19	12	13
05.0110 Russian studies	172	186	169	141	168	145	177	154	169	162
05.0111 Scandinavian studies	26	24	30	25	32	24	32	27	30	29
05.0112 South Asian studies	30	39	17	20	22	33	28	34	23	28
05.0113 Southeast Asian studies	21	28	17	25	15	18	25	14	22	16
05.0114 Western European studies	65	61	54	60	57	60	69	64	73	70
05.0115 Canadian studies	2	4	2	1	2	2	—	1	3	2
05.0116 Balkans studies	—	—	—	—	—	—	—	—	—	—
05.0117 Baltic studies	—	—	—	—	—	—	—	—	—	—
05.0118 Slavic studies	—	—	—	2	2	3	7	7	7	3
05.0119 Caribbean studies	—	—	—	—	—	—	—	—	—	—
05.0120 Ural-Altaic and Central Asian studies	—	—	—	1	10	11	13	12	10	26
05.0121 Commonwealth studies	—	—	—	—	—	—	—	—	1	—
05.0122 Regional studies (U.S., Canadian, foreign)	—	—	—	5	12	7	16	37	35	13
05.0123 Chinese studies	—	—	—	8	19	22	30	25	36	39
05.0124 French studies	—	—	—	10	30	45	56	43	79	53
05.0125 German studies	—	—	—	11	34	35	36	64	45	58
05.0126 Italian studies	—	—	—	18	28	35	42	51	48	55
05.0127 Japanese studies	—	—	—	17	27	44	48	51	65	57
05.0128 Korean studies	—	—	—	—	8	4	4	4	7	2
05.0129 Polish studies	—	—	—	—	—	—	—	—	—	—
05.0130 Spanish and Iberian studies	—	—	—	7	7	13	17	19	18	20
05.0131 Tibetan studies	—	—	—	—	—	—	—	1	—	—
05.0132 Ukraine studies	—	—	—	—	—	—	—	—	—	—
05.0199 Area studies, other	642	621	704	591	605	579	627	637	726	752
11.0401 Information science / studies	9,593	11,505	13,172	14,600	11,149	9,655	9,108	7,856	8,295	8,452
11.0801 Web page, digital / multimedia and information resources design	—	—	—	145	248	486	569	629	751	915

continued

TABLE C.6 Continued

Instructional Program	2000	2001	2002	2003	2004	2005	2006	2007	2008	2009
16.0103 Language interpretation and translation	103	122	145	115	148	151	149	192	178	165
24.0103 Humanities / humanistic studies	3,365	3,716	3,875	3,660	3,564	3,833	3,816	3,496	3,459	3,020
25.0101 Library science / librarianship	4,702	4,712	4,997	5,281	5,997	6,196	6,439	6,765	7,064	7,048
27.0301 Applied mathematics	1,138	1,201	1,191	1,398	1,479	1,660	1,696	1,606	1,637	1,881
27.0303 Computational mathematics	—	—	—	42	109	102	110	83	95	125
27.0501 Statistics, general	1,383	1,377	1,485	1,656	1,931	2,024	2,009	2,135	2,248	2,392
27.0502 Mathematical statistics and probability	—	—	—	6	4	19	16	15	100	110
27.0599 Statistics, other	—	—	—	16	10	12	14	30	24	38
30.2001 International / global studies	—	—	—	1,033	1,637	2,085	2,310	2,494	2,961	3,627
45.0201 Anthropology	8,312	8,226	8,520	8,827	9,022	9,051	9,474	9,552	10,173	10,579
45.0202 Physical anthropology	—	—	—	—	10	16	10	7	1	11
45.0299 Anthropology, other	—	—	—	16	27	28	39	27	52	57
45.1001 Political science and government, general	29,716	29,714	31,011	34,916	37,280	39,869	41,090	41,649	42,097	41,204
45.1101 Sociology	28,278	27,754	27,782	28,680	29,597	30,603	30,660	31,159	31,000	30,979
54.0101 History, general	28,377	27,951	28,900	30,499	32,565	34,476	36,190	37,451	37,555	37,858
54.0102 American history United States	70	71	76	93	83	70	93	130	110	110
54.0103 European history	16	16	24	23	24	28	32	46	40	40
54.0106 Asian history	—	—	—	3	6	2	4	3	5	1
54.0107 Canadian history	—	—	—	—	—	—	—	—	—	—
Total	120,028	121,196	126,116	136,112	140,247	145,912	149,769	151,240	154,140	155,016
Visual analytics										
10.0301 Graphic communications, general	—	—	—	45	149	97	158	215	197	235
10.0303 Prepress / desktop publishing and digital imaging design	—	—	—	—	23	34	28	54	46	71
10.0304 Animation, interactive technology, video graphics and special effects	—	—	—	169	379	563	873	1,232	1,628	1,675
10.0399 Graphic communications, other	—	—	—	70	66	28	15	3	14	69
11.0202 Computer programming special applications	—	—	—	405	212	66	77	68	44	43
11.0401 Information science / studies	9,593	11,505	13,172	14,600	11,149	9,655	9,108	7,856	8,295	8,452
11.0801 Web page, digital / multimedia and information resources design	—	—	—	145	248	486	569	629	751	915
11.0803 Computer graphics	—	—	—	350	375	350	262	563	891	956
14.1301 Engineering science	599	537	655	656	644	663	674	602	643	752
45.0702 Cartography	104	140	93	99	124	135	117	135	126	165
50.0409 Graphic design	—	—	—	1,457	2,955	3,354	3,620	3,734	4,070	4,345
Total	10,296	12,182	13,920	17,996	16,324	15,431	15,501	15,091	16,705	17,678

continued

TABLE C.6 Continued

Instructional Program	2000	2001	2002	2003	2004	2005	2006	2007	2008	2009
Forecasting										
03.0102 Environmental science / studies	5,186	4,942	5,180	3,029	391	—	—	—	—	—
03.0104 Environmental science	—	—	—	1,123	2,410	2,561	2,557	2,680	2,800	3,214
03.0501 Forestry, general	1,115	1,081	974	991	849	1,006	1,084	930	939	971
11.0102 Artificial intelligence and robotics	—	—	—	16	45	47	72	50	97	84
14.0903 Computer software engineering	—	—	—	296	516	805	712	829	1,112	1,326
14.1101 Engineering mechanics	214	252	293	291	280	209	246	196	187	192
14.1301 Engineering science	599	537	655	656	644	663	674	602	643	752
14.3701 Operations research	902	721	619	653	761	839	786	713	809	941
15.1501 Engineering / industrial management	1,297	1,600	1,669	1,687	1,788	1,911	1,911	2,073	2,165	2,661
27.0102 Algebra and number theory	—	—	—	—	—	—	—	—	—	—
27.0301 Applied mathematics	1,138	1,201	1,191	1,398	1,479	1,660	1,696	1,606	1,637	1,881
27.0303 Computational mathematics	—	—	—	42	109	102	110	83	95	125
27.0501 Statistics, general	1,383	1,377	1,485	1,656	1,931	2,024	2,009	2,135	2,248	2,392
27.0502 Mathematical statistics and probability	—	—	—	6	4	19	16	15	100	110
30.0801 Mathematics and computer science	375	371	471	456	437	366	250	230	188	199
40.0202 Astrophysics	103	132	168	123	163	192	190	178	234	192
40.0401 Atmospheric sciences and meteorology, general	733	666	734	767	724	841	765	812	783	830
40.0402 Atmospheric chemistry and climatology	—	—	—	—	—	—	—	—	—	—
40.0403 Atmospheric physics and dynamics	—	—	—	—	6	3	9	8	7	7
40.0404 Meteorology	—	—	—	35	90	176	210	200	270	230
40.0601 Geology / earth science, general	4,358	4,289	4,243	4,219	4,144	4,124	4,290	4,358	4,431	4,681
40.0603 Geophysics and seismology	167	191	197	182	203	203	194	201	217	213
40.0605 Hydrology and water resources science	—	—	—	46	37	51	58	63	65	67
40.0607 Oceanography, chemical and physical	382	372	429	354	398	349	318	393	372	390
40.0801 Physics, general	5,510	5,683	5,785	6,015	6,379	6,741	7,147	7,507	7,573	7,469
40.0805 Plasma and high-temperature physics	—	—	—	—	—	—	—	—	—	—
40.0808 Solid state and low-temperature physics	—	—	—	—	—	—	—	—	—	—
40.0810 Theoretical and mathematical physics	15	4	11	7	9	14	13	20	14	11
45.1001 Political science and government, general	29,716	29,714	31,011	34,916	37,280	39,869	41,090	41,649	42,097	41,204
45.1101 Sociology	28,278	27,754	27,782	28,680	29,597	30,603	30,660	31,159	31,000	30,979
Total	81,471	80,887	82,897	87,644	90,674	95,378	97,067	98,690	100,083	101,121

NOTE: Most instructional programs for which no postsecondary degrees were awarded were new additions to the 2000 Classification of Instructional Programs.
SOURCE: Data from the U.S. Department of Education, National Center for Education Statistics' Integrated Postsecondary Education Data System Completions Survey. Accessed via WebCASPAR.

CITIZENSHIP DATA

TABLE C.7 Number of Degrees Conferred by Year and Citizenship Status Across Highly Relevant Fields of Study

Degree Level	Citizenship	2000	2001	2002	2003	2004	2005	2006	2007	2008	2009
Bachelor's degrees	U.S. citizens and permanent residents	129,403	129,913	135,644	149,223	160,072	166,588	171,232	174,703	175,492	176,978
	Temporary residents	3,018	2,994	3,299	4,018	3,956	4,152	4,430	4,158	3,795	4,084
Master's degrees	U.S. citizens and permanent residents	22,682	22,653	23,438	25,016	27,824	28,987	30,533	30,663	32,087	32,746
	Temporary residents	4,673	5,103	5,543	6,500	6,942	6,656	6,337	5,864	6,990	7,991
Doctorate degrees	U.S. citizens and permanent residents	4,786	4,504	4,283	4,396	4,352	4,228	4,500	4,664	4,819	5,155
	Temporary residents	1,817	1,857	1,913	1,920	2,066	2,341	2,512	2,613	2,695	2,834
TOTAL	U.S. citizens and permanent residents	156,871	157,070	163,365	178,635	192,248	199,803	206,265	210,030	212,398	214,879
	Temporary residents	9,508	9,954	10,755	12,438	12,964	13,149	13,279	12,635	13,480	14,909

SOURCE: Data from the U.S. Department of Education, National Center for Education Statistics' Integrated Postsecondary Education Data System Completions Survey. Accessed via WebCASPAR.

TABLE C.8 Percentage of Degrees Conferred by Year to U.S. Citizens and Permanent Residents Across Highly Relevant Fields of Study

Degree Level	Percent of Degrees									
	2000	2001	2002	2003	2004	2005	2006	2007	2008	2009
Bachelor's degrees	98	98	98	97	98	98	97	98	98	98
Master's degrees	83	82	81	79	80	81	83	84	82	80
Doctorate degrees	72	71	69	70	68	64	64	64	64	65
TOTAL	94	94	94	93	94	94	94	94	94	94

SOURCE: Data from the U.S. Department of Education, National Center for Education Statistics' Integrated Postsecondary Education Data System Completions Survey. Accessed via WebCASPAR.

TABLE C.9 Percentage of Degrees Conferred by Year to U.S. Citizens and Permanent Residents in Fields of Study That Are Highly Relevant to the Core and Emerging Areas

Area	Degree	2000	2001	2002	2003	2004	2005	2006	2007	2008	2009
Geodesy and geophysics	Bachelor's	92	91	93	93	93	93	95	94	94	94
	Master's	57	58	57	63	65	66	72	78	76	75
	Doctorate	58	50	43	43	44	41	42	46	43	51
	Total	78	78	79	81	82	82	85	86	86	86
Photogrammetry	Bachelor's	—	—	—	—	100	96	100	94	100	100
	Master's	—	—	—	—	50	70	50	60	75	83
	Doctorate	—	—	—	—	100	0	0	33	—	0
	Total	—	—	—	—	76	78	88	86	95	93
Remote sensing	Bachelor's	97	97	97	97	97	96	96	97	97	97
	Master's	77	75	74	73	72	74	76	76	76	74
	Doctorate	61	59	55	58	56	53	54	56	54	56
	Total	90	90	90	89	88	88	89	89	89	88
Cartographic science	Bachelor's	97	97	97	95	96	96	97	97	97	97
	Master's	83	82	82	78	79	84	82	82	85	87
	Doctorate	74	72	72	71	76	67	70	70	65	67
	Total	95	95	94	93	94	94	95	95	95	95
GIS and geospatial analysis	Bachelor's	98	98	97	98	98	97	97	97	98	98
	Master's	83	83	82	79	79	82	83	83	82	81
	Doctorate	71	69	67	64	66	60	61	61	59	58
	Total	92	92	91	90	90	90	91	91	90	90
GEOINT fusion	Bachelor's	95	95	95	95	96	96	95	95	97	97
	Master's	77	75	74	71	76	79	80	83	75	73
	Doctorate	66	66	64	70	67	64	61	65	65	66
	Total	91	90	90	89	91	92	91	92	90	89
Crowdsourcing	Bachelor's	93	93	91	92	95	93	91	92	95	96
	Master's	54	50	45	48	57	51	52	53	47	46
	Doctorate	56	44	35	44	35	33	29	36	33	39
	Total	65	61	56	70	82	81	80	80	72	72
Human geography	Bachelor's	98	98	98	98	98	98	98	98	98	98
	Master's	87	86	84	83	84	86	87	88	87	86
	Doctorate	84	82	82	82	80	78	77	76	75	75
	Total	96	96	96	95	96	96	96	97	96	96
Visual analytics	Bachelor's	93	94	94	93	94	95	95	96	96	96
	Master's	69	68	67	66	70	74	78	81	75	73
	Doctorate	59	53	61	63	57	50	46	54	46	52
	Total	87	87	86	87	88	90	91	92	91	90
Forecasting	Bachelor's	98	98	98	98	98	98	98	98	98	98
	Master's	79	77	76	74	73	73	74	75	73	70
	Doctorate	69	67	66	66	63	59	59	59	59	60
	Total	94	94	94	94	93	93	93	94	93	93

SOURCE: Data from the U.S. Department of Education, National Center for Education Statistics' Integrated Postsecondary Education Data System Completions Survey. Accessed via WebCASPAR.

TABLE C.10 Total Number of Degrees Conferred to U.S. Citizens and Permanent Residents by Year in Fields of Study That Are Highly Relevant to the Core and Emerging Areas

Instructional Program	2000	2001	2002	2003	2004	2005	2006	2007	2008	2009
Geodesy and geophysics										
14.0201 Aerospace, aeronautical and astronautical engineering	1,610	1,811	2,097	2,378	2,796	2,960	3,529	3,618	3,790	3,934
14.1201 Engineering physics	288	280	297	322	370	392	421	475	471	454
14.1301 Engineering science	412	373	459	491	475	460	475	466	482	578
14.3801 Surveying engineering	—	—	—	—	13	36	23	36	42	26
40.0603 Geophysics and seismology	118	131	133	129	141	139	137	134	143	138
Total	2,428	2,595	2,986	3,320	3,795	3,987	4,585	4,729	4,928	5,130
Photogrammetry										
14.3801 Surveying engineering	—	—	—	—	13	36	23	36	42	26
Total	—	—	—	—	13	36	23	36	42	26
Remote sensing										
03.0206 Land use planning and management / development	—	—	—	27	96	82	88	107	92	128
03.0506 Forest management / forest resources management	253	206	194	179	155	146	125	129	141	141
04.0301 City / urban, community and regional planning	1,654	1,553	1,670	1,857	2,021	2,272	2,440	2,523	2,482	2,652
11.0203 Computer programming, vendor / product certification	—	—	—	—	—	—	—	—	—	—
14.3801 Surveying engineering	—	—	—	—	13	36	23	36	42	26
14.3901 Geological / geophysical engineering	217	175	162	156	147	128	123	168	166	195
15.1102 Survey technology / surveying	188	230	195	196	187	195	195	230	243	242
27.0101 Mathematics, general	11,586	11,139	11,873	12,176	13,010	14,017	14,689	15,132	15,240	15,300
27.0105 Topology and foundations	—	—	—	—	—	—	—	—	—	—
27.0501 Statistics, general	896	836	834	905	1,003	1,098	1,064	1,154	1,204	1,302
27.0502 Mathematical statistics and probability	—	—	—	5	3	12	13	14	47	44
29.0101 Military technologies	154	20	3	34	10	244	316	59	351	428
30.0801 Mathematics and computer science	333	324	412	400	386	300	211	201	145	155
40.0101 Physical sciences	369	348	336	322	257	334	307	285	318	340
40.0403 Atmospheric physics and dynamics	—	—	—	—	6	2	9	8	7	7
40.0601 Geology / earth science, general	4,149	4,067	3,998	3,998	3,900	3,861	4,019	4,092	4,166	4,371
40.0603 Geophysics and seismology	118	131	133	129	141	139	137	134	143	138
40.0807 Optics / optical sciences	33	43	40	118	118	119	129	126	143	119
45.0701 Geography	4,704	4,707	4,649	4,755	5,016	5,013	4,956	5,138	5,233	5,404
45.0702 Cartography	100	137	84	96	118	123	111	121	112	155
45.0799 Geography, other	—	—	—	46	73	114	115	161	219	174
Total	24,754	23,916	24,583	25,399	26,660	28,235	29,070	29,818	30,494	31,321

continued

TABLE C.10 Continued

Instructional Program	2000	2001	2002	2003	2004	2005	2006	2007	2008	2009
Cartographic science										
10.0303 Prepress / desktop publishing and digital imaging design	—	—	—	—	23	34	28	53	46	71
10.0304 Animation, interactive technology, video graphics and special effects	—	—	—	148	360	516	825	1,159	1,555	1,616
11.0201 Computer programming / programmer, general	474	471	702	835	723	517	432	433	389	354
11.0202 Computer programming special applications	—	—	—	312	174	65	67	55	32	38
11.0203 Computer programming, vendor / product certification	—	—	—	—	—	—	—	—	—	—
14.3801 Surveying engineering	—	—	—	—	13	36	23	36	42	26
15.1102 Survey technology / surveying	188	230	195	196	187	195	195	230	243	242
27.0104 Geometry / geometric analysis	—	—	—	—	—	—	—	—	—	—
45.0701 Geography	4,704	4,707	4,649	4,755	5,016	5,013	4,956	5,138	5,233	5,404
45.0702 Cartography	100	137	84	96	118	123	111	121	112	155
50.0401 Design and visual communications, general	1,311	1,376	1,667	2,200	2,868	2,354	2,543	2,364	2,318	2,074
50.0409 Graphic design	—	—	—	1,344	2,739	3,178	3,411	3,530	3,825	4,104
Total	6,777	6,921	7,297	9,886	12,221	12,031	12,591	13,119	13,795	14,084
GIS and geospatial analysis										
03.0506 Forest management / forest resources management	253	206	194	179	155	146	125	129	141	141
04.0301 City / urban, community and regional planning	1,654	1,553	1,670	1,857	2,021	2,272	2,440	2,523	2,482	2,652
14.3701 Operations research	785	622	513	422	533	617	543	474	537	584
45.0701 Geography	4,704	4,707	4,649	4,755	5,016	5,013	4,956	5,138	5,233	5,404
45.0702 Cartography	100	137	84	96	118	123	111	121	112	155
Total	7,496	7,225	7,110	7,309	7,843	8,171	8,175	8,385	8,505	8,936
GEOINT fusion										
03.0102 Environmental science / studies	4,985	4,730	4,936	2,884	376	—	—	—	—	—
03.0103 Environmental studies	—	—	—	1,289	2,976	3,146	3,262	3,176	3,466	3,469
03.0104 Environmental science	—	—	—	1,057	2,256	2,414	2,377	2,539	2,657	3,051
03.0501 Forestry, general	1,035	996	890	906	772	929	1,001	851	862	897
03.0506 Forest management / forest resources management	253	206	194	179	155	146	125	129	141	141
11.0102 Artificial intelligence and robotics	—	—	—	6	22	24	42	30	49	43
11.0103 Information technology	—	—	—	1,034	5,606	5,153	5,249	5,252	2,867	3,120
11.0401 Information science / studies	8,406	10,065	11,492	12,665	9,745	8,680	8,202	7,200	7,311	7,271
11.0802 Data modeling / warehousing and database administration	—	—	—	201	147	46	77	65	45	54
14.3801 Surveying engineering	—	—	—	—	13	36	23	36	42	26
15.1102 Survey technology / surveying	188	230	195	196	187	195	195	230	243	242
27.0502 Mathematical statistics and probability	—	—	—	5	3	12	13	14	47	44
40.0401 Atmospheric sciences and meteorology, general	657	610	675	694	652	748	676	746	718	770
45.0702 Cartography	100	137	84	96	118	123	111	121	112	155
Total	15,624	16,974	18,466	21,212	23,028	21,652	21,353	20,389	18,560	19,283

continued

TABLE C.10 Continued

Instructional Program	2000	2001	2002	2003	2004	2005	2006	2007	2008	2009
Crowdsourcing										
11.0103 Information technology	—	—	—	1,034	5,606	5,153	5,249	5,252	2,867	3,120
11.0199 Computer and information science, other	—	—	—	152	127	126	112	111	104	148
27.0501 Statistics, general	896	836	834	905	1,003	1,098	1,064	1,154	1,204	1,302
27.0502 Mathematical statistics and probability	—	—	—	5	3	12	13	14	47	44
27.0599 Statistics, other	—	—	—	13	10	6	11	22	20	30
Total	896	836	834	2,109	6,749	6,395	6,449	6,553	4,242	4,644
Human geography										
05.0101 African studies	61	38	39	60	58	48	68	53	90	117
05.0102 American / United States studies / civilization	1,778	1,877	1,899	1,952	1,939	1,986	1,988	1,950	1,948	1,890
05.0103 Asian studies / civilization	510	465	472	445	492	553	638	673	735	772
05.0104 East Asian studies	437	344	365	348	388	377	438	469	519	529
05.0105 Central / middle and eastern European studies	8	27	39	32	13	19	16	15	29	31
05.0106 European studies / civilization	97	121	134	123	111	136	185	172	165	188
05.0107 Latin American studies	646	589	577	679	717	758	752	789	778	771
05.0108 Near and Middle Eastern studies	195	158	150	214	269	289	299	277	318	372
05.0109 Pacific area / Pacific Rim studies	10	8	11	9	13	23	19	19	7	11
05.0110 Russian studies	160	175	159	126	154	132	162	139	162	152
05.0111 Scandinavian studies	26	23	29	25	30	24	31	26	28	29
05.0112 South Asian studies	26	29	17	18	16	27	22	31	22	26
05.0113 Southeast Asian studies	21	22	16	23	9	9	18	11	13	12
05.0114 Western European studies	56	55	46	54	44	53	59	53	63	65
05.0115 Canadian studies	2	4	2	1	2	2	—	1	3	2
05.0116 Balkans studies	—	—	—	—	—	—	—	—	—	—
05.0117 Baltic studies	—	—	—	—	—	—	—	—	—	—
05.0118 Slavic studies	—	—	—	2	2	3	7	7	6	3
05.0119 Caribbean studies	—	—	—	—	—	—	—	—	—	—
05.0120 Ural-Altaic and Central Asian studies	—	—	—	1	7	6	9	12	9	21
05.0121 Commonwealth studies	—	—	—	—	—	—	—	—	1	—
05.0122 Regional studies (U.S., Canadian, foreign)	—	—	—	5	12	7	16	37	34	13
05.0123 Chinese studies	—	—	—	6	16	19	25	21	36	37
05.0124 French studies	—	—	—	9	24	41	50	36	71	48
05.0125 German studies	—	—	—	9	30	33	33	58	41	55
05.0126 Italian studies	—	—	—	16	27	35	42	48	45	51
05.0127 Japanese studies	—	—	—	17	27	41	45	46	64	54
05.0128 Korean studies	—	—	—	—	5	1	3	3	4	1
05.0129 Polish studies	—	—	—	—	—	—	—	—	—	—
05.0130 Spanish and Iberian studies	—	—	—	7	7	13	17	19	18	20
05.0131 Tibetan studies	—	—	—	—	—	—	—	1	—	—
05.0132 Ukraine studies	—	—	—	—	—	—	—	—	—	—
05.0199 Area studies, other	599	559	653	556	575	546	600	611	702	708
11.0401 Information science / studies	8,406	10,065	11,492	12,665	9,745	8,680	8,202	7,200	7,311	7,271
11.0801 Web page, digital / multimedia and information resources design	—	—	—	129	232	472	538	614	733	847

continued

TABLE C.10 Continued

Instructional Program	2000	2001	2002	2003	2004	2005	2006	2007	2008	2009
16.0103 Language interpretation and translation	45	55	76	58	77	87	95	121	118	114
24.0103 Humanities / humanistic studies	3,233	3,598	3,716	3,538	3,454	3,729	3,724	3,422	3,354	2,933
25.0101 Library science / librarianship	4,545	4,569	4,831	5,104	5,793	6,022	6,282	6,628	6,937	6,928
27.0301 Applied mathematics	946	984	986	1,124	1,148	1,317	1,347	1,307	1,316	1,424
27.0303 Computational mathematics	—	—	—	39	90	82	93	64	67	94
27.0501 Statistics, general	896	836	834	905	1,003	1,098	1,064	1,154	1,204	1,302
27.0502 Mathematical statistics and probability	—	—	—	5	3	12	13	14	47	44
27.0599 Statistics, other	—	—	—	13	10	6	11	22	20	30
30.2001 International / global studies	—	—	—	943	1,507	1,893	2,130	2,298	2,765	3,362
45.0201 Anthropology	8,053	7,957	8,235	8,590	8,750	8,802	9,209	9,258	9,894	10,279
45.0202 Physical anthropology	—	—	—	—	9	16	9	7	1	10
45.0299 Anthropology, other	—	—	—	15	27	27	38	26	51	55
45.1001 Political science and government, general	28,778	28,743	29,997	33,768	36,200	38,758	39,922	40,517	41,009	40,047
45.1101 Sociology	27,709	27,268	27,236	28,055	28,977	30,018	30,012	30,494	30,386	30,333
54.0101 History, general	27,997	27,602	28,519	30,129	32,219	34,069	35,777	37,067	37,144	37,413
54.0102 American history United States	70	69	74	92	82	69	88	130	108	108
54.0103 European history	14	15	21	20	19	27	32	46	40	39
54.0106 Asian history	—	—	—	—	5	2	4	2	3	1
54.0107 Canadian history	—	—	—	—	—	—	—	—	—	—
Total	115,324	116,255	120,625	129,929	134,337	140,367	144,132	145,968	148,419	148,612
Visual analytics										
10.0301 Graphic communications, general	—	—	—	45	148	97	155	206	187	232
10.0303 Prepress / desktop publishing and digital imaging design	—	—	—	—	23	34	28	53	46	71
10.0304 Animation, interactive technology, video graphics and special effects	—	—	—	148	360	516	825	1,159	1,555	1,616
10.0399 Graphic communications, other	—	—	—	69	62	27	15	3	14	62
11.0202 Computer programming special applications	—	—	—	312	174	65	67	55	32	38
11.0401 Information science / studies	8,406	10,065	11,492	12,665	9,745	8,680	8,202	7,200	7,311	7,271
11.0801 Web page, digital / multimedia and information resources design	—	—	—	129	232	472	538	614	733	847
11.0803 Computer graphics	—	—	—	312	323	307	242	545	835	858
14.1301 Engineering science	412	373	459	491	475	460	475	466	482	578
45.0702 Cartography	100	137	84	96	118	123	111	121	112	155
50.0409 Graphic design	—	—	—	1,344	2,739	3,178	3,411	3,530	3,825	4,104
Total	8,918	10,575	12,035	15,611	14,399	13,959	14,069	13,952	15,132	15,832

continued

TABLE C.10 Continued

Instructional Program	2000	2001	2002	2003	2004	2005	2006	2007	2008	2009
Forecasting										
03.0102 Environmental science / studies	4,985	4,730	4,936	2,884	376	—	—	—	—	—
03.0104 Environmental science	—	—	—	1,057	2,256	2,414	2,377	2,539	2,657	3,051
03.0501 Forestry, general	1,035	996	890	906	772	929	1,001	851	862	897
11.0102 Artificial intelligence and robotics	—	—	—	6	22	24	42	30	49	43
14.0903 Computer software engineering	—	—	—	228	365	543	512	644	791	922
14.1101 Engineering mechanics	142	178	214	214	199	145	193	157	137	138
14.1301 Engineering science	412	373	459	491	475	460	475	466	482	578
14.3701 Operations research	785	622	513	422	533	617	543	474	537	584
15.1501 Engineering / industrial management	1,104	1,312	1,328	1,345	1,416	1,535	1,567	1,682	1,652	1,926
27.0102 Algebra and number theory	—	—	—	—	—	—	—	—	—	—
27.0301 Applied mathematics	946	984	986	1,124	1,148	1,317	1,347	1,307	1,316	1,424
27.0303 Computational mathematics	—	—	—	39	90	82	93	64	67	94
27.0501 Statistics, general	896	836	834	905	1,003	1,098	1,064	1,154	1,204	1,302
27.0502 Mathematical statistics and probability	—	—	—	5	3	12	13	14	47	44
30.0801 Mathematics and computer science	333	324	412	400	386	300	211	201	145	155
40.0202 Astrophysics	93	110	150	109	143	170	177	162	207	177
40.0401 Atmospheric sciences and meteorology, general	657	610	675	694	652	748	676	746	718	770
40.0402 Atmospheric chemistry and climatology	—	—	—	—	—	—	—	—	—	—
40.0403 Atmospheric physics and dynamics	—	—	—	—	6	2	9	8	7	7
40.0404 Meteorology	—	—	—	29	83	170	202	183	255	217
40.0601 Geology / earth science, general	4,149	4,067	3,998	3,998	3,900	3,861	4,019	4,092	4,166	4,371
40.0603 Geophysics and seismology	118	131	133	129	141	139	137	134	143	138
40.0605 Hydrology and water resources science	—	—	—	42	35	45	50	59	61	63
40.0607 Oceanography, chemical and physical	335	329	373	316	344	286	260	325	319	326
40.0801 Physics, general	4,443	4,533	4,680	4,826	5,136	5,293	5,684	6,030	6,051	6,001
40.0805 Plasma and high-temperature physics	—	—	—	—	—	—	—	—	—	—
40.0808 Solid state and low-temperature physics	—	—	—	—	—	—	—	—	—	—
40.0810 Theoretical and mathematical physics	15	4	10	7	8	14	13	19	14	11
45.1001 Political science and government, general	28,778	28,743	29,997	33,768	36,200	38,758	39,922	40,517	41,009	40,047
45.1101 Sociology	27,709	27,268	27,236	28,055	28,977	30,018	30,012	30,494	30,386	30,333
Total	76,935	76,150	77,824	81,999	84,669	88,980	90,599	92,352	93,282	93,619

SOURCE: Data from the U.S. Department of Education, National Center for Education Statistics' Integrated Postsecondary Education Data System Completions Survey. Accessed via WebCASPAR.

Appendix D

Data on Occupations

TABLE D.1 2010 Bureau of Labor Statistics Codes and Descriptions of 36 Occupations That Are Relevant to NGA

Code	Title	Description
15-1111	Computer and information research scientists	Conduct research into fundamental computer and information science as theorists, designers, or inventors. Develop solutions to problems in the field of computer hardware and software
15-1121	Computer systems analysts	Analyze science, engineering, business, and other data processing problems to implement and improve computer systems. Analyze user requirements, procedures, and problems to automate or improve existing systems and review computer system capabilities, workflow, and scheduling limitations. May analyze or recommend commercially available software
15-1131	Computer programmers	Create, modify, and test the code, forms, and script that allow computer applications to run. Work from specifications drawn up by software developers or other individuals. May assist software developers by analyzing user needs and designing software solutions. May develop and write computer programs to store, locate, and retrieve specific documents, data, and information
15-1132	Software developers, applications	Develop, create, and modify general computer applications software or specialized utility programs. Analyze user needs and develop software solutions. Design software or customize software for client use with the aim of optimizing operational efficiency. May analyze and design databases within an application area, working individually or coordinating database development as part of a team. May supervise computer programmers
15-1799	Computer occupations, all other	All computer specialists not listed separately (e.g., computer laboratory technician)
15-2021	Mathematicians	Conduct research in fundamental mathematics or in application of mathematical techniques to science, management, and other fields. Solve problems in various fields using mathematical methods
15-2031	Operations research analysts	Formulate and apply mathematical modeling and other optimizing methods to develop and interpret information that assists management with decision making, policy formulation, or other managerial functions. May collect and analyze data and develop decision support software, service, or products. May develop and supply optimal time, cost, or logistics networks for program evaluation, review, or implementation
15-2041	Statisticians	Develop or apply mathematical or statistical theory and methods to collect, organize, interpret, and summarize numerical data to provide usable information. May specialize in fields such as biostatistics, agricultural statistics, business statistics, or economic statistics. Includes mathematical and survey statisticians. Excludes survey researchers
15-2091	Mathematical technicians	Apply standardized mathematical formulas, principles, and methodology to technological problems in engineering and physical sciences in relation to specific industrial and research objectives, processes, equipment, and products
15-2099	Mathematical science occupations, all other	All mathematical scientists not listed separately (e.g., harmonic analyst)

continued

TABLE D.1 Continued

Code	Title	Description
17-1021	Cartographers and photogrammetrists	Collect, analyze, and interpret geographic information provided by geodetic surveys, aerial photographs, and satellite data. Research, study, and prepare maps and other spatial data in digital or graphic form for legal, social, political, educational, and design purposes. May work with Geographic Information Systems (GIS). May design and evaluate algorithms, data structures, and user interfaces for GIS and mapping systems
17-1022	Surveyors	Make exact measurements and determine property boundaries. Provide data relevant to the shape, contour, gravitation, location, elevation, or dimension of land or land features on or near the earth's surface for engineering, mapmaking, mining, land evaluation, construction, and other purposes
17-2011	Aerospace engineers	Perform engineering duties in designing, constructing, and testing aircraft, missiles, and spacecraft. May conduct basic and applied research to evaluate adaptability of materials and equipment to aircraft design and manufacture. May recommend improvements in testing equipment and techniques
17-2061	Computer hardware engineers	Research, design, develop, or test computer or computer-related equipment for commercial, industrial, military, or scientific use. May supervise the manufacturing and installation of computer or computer-related equipment and components. Excludes software developers, applications and software developers, systems software
17-2071	Electrical engineers	Research, design, develop, test, or supervise the manufacturing and installation of electrical equipment, components, or systems for commercial, industrial, military, or scientific use. Excludes computer hardware engineers
17-2199	Engineers, all other	All engineers not listed separately (e.g., photonics engineer, optical engineer)
17-3031	Surveying and mapping technicians	Perform surveying and mapping duties, usually under the direction of an engineer, surveyor, cartographer, or photogrammetrist to obtain data used for construction, mapmaking, boundary location, mining, or other purposes. May calculate mapmaking information and create maps from source data, such as surveying notes, aerial photography, satellite data, or other maps to show topographical features, political boundaries, and other features. May verify accuracy and completeness of maps. Excludes surveyors; cartographers and photogrammetrists; and geoscientists, except hydrologists and geographers
19-2011	Astronomers	Observe, research, and interpret astronomical phenomena to increase basic knowledge or apply such information to practical problems
19-2012	Physicists	Conduct research into physical phenomena, develop theories on the basis of observation and experiments, and devise methods to apply physical laws and theories. Excludes biochemists and biophysicists
19-2021	Atmospheric and space scientists	Investigate atmospheric phenomena and interpret meteorological data, gathered by surface and air stations, satellites, and radar to prepare reports and forecasts for public and other uses. Includes weather analysts and forecasters whose functions require the detailed knowledge of meteorology
19-2041	Environmental scientists and specialists, including health	Conduct research or perform investigation for the purpose of identifying, abating, or eliminating sources of pollutants or hazards that affect either the environment or the health of the population. Using knowledge of various scientific disciplines, may collect, synthesize, study, report, and recommend action based on data derived from measurements or observations of air, food, soil, water, and other sources. Excludes zoologists and wildlife biologists, conservation scientists, forest and conservation technicians, fish and game wardens, and forest and conservation workers
19-2042	Geoscientists, except hydrologists and geographers	Study the composition, structure, and other physical aspects of the Earth. May use geological, physics, and mathematics knowledge in exploration for oil, gas, minerals, or underground water; or in waste disposal, land reclamation, or other environmental problems. May study the Earth's internal composition, atmospheres, oceans, and its magnetic, electrical, and gravitational forces. Includes mineralogists, crystallographers, paleontologists, stratigraphers, geodesists, and seismologists
19-2043	Hydrologists	Research the distribution, circulation, and physical properties of underground and surface waters; and study the form and intensity of precipitation, its rate of infiltration into the soil, movement through the Earth, and its return to the ocean and atmosphere
19-2099	Physical scientists, all other	All physical scientists not listed separately
19-3041	Sociologists	Study human society and social behavior by examining the groups and social institutions that people form, as well as various social, religious, political, and business organizations. May study the behavior and interaction of groups, trace their origin and growth, and analyze the influence of group activities on individual members
19-3051	Urban and regional planners	Develop comprehensive plans and programs for use of land and physical facilities of local jurisdictions, such as towns, cities, counties, and metropolitan areas

continued

TABLE D.1 Continued

Code	Title	Description
19-3091	Anthropologists and archeologists	Study the origin, development, and behavior of human beings. May study the way of life, language, or physical characteristics of people in various parts of the world. May engage in systematic recovery and examination of material evidence, such as tools or pottery remaining from past human cultures, in order to determine the history, customs, and living habits of earlier civilizations
19-3092	Geographers	Study the nature and use of areas of the Earth's surface, relating and interpreting interactions of physical and cultural phenomena. Conduct research on physical aspects of a region, including land forms, climates, soils, plants, and animals, and conduct research on the spatial implications of human activities within a given area, including social characteristics, economic activities, and political organization, as well as researching interdependence between regions at scales ranging from local to global
19-3093	Historians	Research, analyze, record, and interpret the past as recorded in sources, such as government and institutional records, newspapers and other periodicals, photographs, interviews, films, electronic media, and unpublished manuscripts, such as personal diaries and letters
19-3094	Political scientists	Study the origin, development, and operation of political systems. May study topics, such as public opinion, political decision making, and ideology. May analyze the structure and operation of governments, as well as various political entities. May conduct public opinion surveys, analyze election results, or analyze public documents. Excludes survey researchers
19-3099	Social scientists and related workers, all other	All social scientists and related workers not listed separately
19-4093	Forest and conservation technicians	Provide technical assistance regarding the conservation of soil, water, forests, or related natural resources. May compile data pertaining to size, content, condition, and other characteristics of forest tracts, under the direction of foresters; or train and lead forest workers in forest propagation, fire prevention and suppression. May assist conservation scientists in managing, improving, and protecting rangelands and wildlife habitats. Excludes conservation scientists and foresters
19-4099	Life, physical, and social science technicians, all other	All life, physical, and social science technicians not listed separately
25-4021	Librarians	Administer libraries and perform related library services. Work in a variety of settings, including public libraries, educational institutions, museums, corporations, government agencies, law firms, nonprofit organizations, and healthcare providers. Tasks may include selecting, acquiring, cataloguing, classifying, circulating, and maintaining library materials, and furnishing reference, bibliographical, and readers' advisory services. May perform in-depth, strategic research, and synthesize, analyze, edit, and filter information. May set up or work with databases and information systems to catalogue and access information
27-1014	Multi-media artists and animators	Create special effects, animation, or other visual images using film, video, computers, or other electronic tools and media for use in products or creations, such as computer games, movies, music videos, and commercials
43-9111	Statistical assistants	Compile and compute data according to statistical formulas for use in statistical studies. May perform actuarial computations and compile charts and graphs for use by actuaries

SOURCE: Bureau of Labor Statistics Standard Occupational Classification, 2010 version, <http://bls.gov/oes/>.

TABLE D.2 Employment and Salary of NGA-Relevant Occupations

Occupation	Number of Jobs (2010)[a]			Mean Annual Salary (2010)		
	All Sectors[b]	Private Sector	Federal Government	All Sectors	Private Sector	Federal Government
15-1111 Computer and information research scientists	24,900	18,180	6,080	$103,150	$104,110	$102,070
15-1121 Computer systems analysts	495,800	442,120	660	$81,250	$82,800	$79,750
15-1131 Computer programmers	333,620	308,360	60	$74,900	$75,840	$88,790
15-1132 Software developers, applications	499,280	476,080	—	$90,410	$91,290	—
15-1799 Computer occupations, all other	183,110	97,910	68,600	$79,790	$75,050	$90,480
15-2021 Mathematicians	2,830	1,420	1,020	$100,260	$103,080	$106,950
15-2031 Operations research analysts	62,210	50,070	4,670	$76,980	$77,250	$105,840
15-2041 Statisticians	22,830	13,530	4,650	$76,070	$77,730	$93,770
15-2091 Mathematical technicians	960	440	70	$49,170	$60,240	$34,030
15-2099 Mathematical science occupations, all other	1,290	870	210	$70,760	$78,590	$52,370
17-1021 Cartographers and photogrammetrists	11,670	7,280	670	$60,970	$61,790	$82,980
17-1022 Surveyors	43,950	38,680	480	$58,140	$56,860	$82,230
17-2011 Aerospace engineers	78,450	68,720	9,220	$99,000	$97,680	$110,780
17-2061 Computer hardware engineers	66,960	62,100	4,430	$101,600	$101,790	$102,200
17-2071 Electrical engineers	148,770	140,260	4,260	$87,770	$88,040	$89,410
17-2199 Engineers, all other	139,610	105,620	25,490	$91,770	$88,800	$108,690
17-3031 Surveying and mapping technicians	53,870	42,620	1,340	$40,370	$39,210	$47,350
19-2011 Astronomers	1,840	950	440	$93,340	$86,520	$132,010
19-2012 Physicists	16,860	11,680	3,210	$112,020	$117,050	$113,470
19-2021 Atmospheric and space scientists	8,640	4,210	3,010	$88,010	$83,250	$95,760
19-2041 Environmental scientists and specialists, including health	81,690	39,960	5,850	$67,810	$70,950	$95,680
19-2042 Geoscientists, except hydrologists and geographers	30,830	23,870	2,460	$93,380	$97,890	$95,580
19-2043 Hydrologists	6,910	3,390	2,000	$79,280	$82,070	$82,900
19-2099 Physical scientists, all other	24,690	11,650	8,460	$95,780	$100,030	$104,620
19-3041 Sociologists	3,710	2,400	—	$80,130	$84,350	—
19-3051 Urban and regional planners	38,830	8,880	750	$66,020	$73,110	$88,740
19-3091 Anthropologists and archeologists	5,100	3,060	1,360	$58,040	$53,130	$71,940
19-3092 Geographers	1,300	340	770	$72,890	$72,200	$76,770
19-3093 Historians	3,320	1,140	760	$57,840	$58,080	$88,130
19-3094 Political scientists	4,470	1,360	2,610	$107,930	$109,990	$115,890
19-3099 Social scientists and related workers, all other	28,420	8,160	15,260	$77,890	$78,240	$83,170
19-4093 Forest and conservation technicians	32,290	1,060	25,070	$36,860	$37,420	$36,680
19-4099 Life, physical, and social science technicians, all other	55,360	32,840	7,660	$45,980	$46,430	$51,890
25-4021 Librarians	148,240	30,020	1,720	$56,360	$55,040	$80,500
27-1014 Multi-media artists and animators	26,560	25,760	—	$63,440	$63,750	—
43-9111 Statistical assistants	15,490	7,280	2,190	$37,090	$41,850	$34,340
TOTAL	2,704,660	2,092,270	215,490			

[a] Estimates include workers who are paid a wage or salary. They do not include the self-employed, owners and partners in unincorporated firms, household workers, or unpaid family workers.

[b] Sectors include federal government, state government, local government, and private companies.

SOURCE: Bureau of Labor Statistics, occupational employment statistics from May 2010, <www.bls.gov/oes/>.

TABLE D.3 Percent of U.S. Citizens Employed in NGA-Relevant Occupations in 2010 That Are Serving or Have Served in the Military

Occupation Title	Percent Serving or Served in Military
Computer and information research scientists	9.7
Computer systems analysts	10.6
Computer programmers	10.2
Software developers, applications and systems software	9.7
Computer occupations, all other	15.6
Operations research analysts	18.3
Miscellaneous mathematical science occupations, including mathematicians and statisticians	6.1
Surveyors, cartographers, and photogrammetrists	11.1
Aerospace engineers	15.1
Computer hardware engineers	9.8
Electrical and electronics engineers	15.6
Surveying and mapping technicians	16.2
Astronomers and physicists	11.3
Atmospheric and space scientists	39.5
Environmental scientists and geoscientists	9.4
Physical scientists, all other	5.4
Miscellaneous social scientists, including survey researchers and sociologists	15.7
Miscellaneous life, physical, and social science technicians, including social science research assistants	8.0
Librarians	2.3
Artists and related workers	6.0
Statistical assistants	7.3

SOURCE: Data from the U.S. Census Bureau, American Community Survey, 2010 Public Use Microdata Sample. Includes those employed as of the survey reference period. Occupation titles are based on the 2010 Census occupational classification system (and are consistent with the 2010 Standard Occupation Code system).

TABLE D.4 Annual Average Unemployment Rates for Wage and Salary Workers in NGA-Relevant Occupations

Occupation Title[a]	Unemployment Rate[b]				
	2006	2007	2008	2009	2010
Computer scientists and systems analysts	2.5	2.1	2.2	6.0	5.9
Computer programmers	2.4	2.2	3.5	4.8	5.8
Computer software engineers	2.1	1.7	1.7	4.2	4.5
Mathematicians	NA	NA	NA	NA	NA
Operations research analysts	1.5	1.5	1.5	2.8	1.9
Statisticians	1.8	NA	2.6	3.1	0.8
Surveyors, cartographers, and photogrammetrists	1.6	4.3	3.2	7.4	2.2
Aerospace engineers	1.3	0.6	1.3	1.6	4.7
Computer hardware engineers	1.2	2.6	1.5	5.4	4.3
Electrical and electronics engineers	1.7	0.9	2.3	6.6	5.5
Surveying and mapping technicians	6.9	3.5	6.9	11.2	15.0
Astronomers and physicists	2.3	5.1	5.0	NA	3.1
Atmospheric and space scientists	17.0	1.0	NA	NA	NA
Environmental scientists and geoscientists	1.3	2.8	1.3	4.7	2.3
Physical scientists, all other	0.9	1.0	1.9	2.7	2.7
Sociologists	NA	0.7	NA	NA	6.7
Urban and regional planners	2.8	0.8	0.9	5.0	1.2
Miscellaneous social scientists and related workers	5.8	6.9	4.3	6.0	5.1
Other life, physical, and social science technicians	2.0	1.2	1.4	7.4	6.5
Librarians	1.6	1.0	3.4	4.1	2.5
Artists and related workers	5.4	5.8	8.1	7.8	13.8
Statistical assistants	2.3	2.9	2.6	10.7	6.8
Management, professional, and related occupations[c]	2.1	2.1	2.7	4.7	4.8

NOTE: NA = not available

[a] Occupation titles are based on the 2002 Census occupational classification system (and are consistent with the 2000 Standard Occupation Code system).

[b] The unemployment rate is the percentage of the labor force that is unemployed. It is calculated by dividing the number of people who are unemployed (i.e., people without jobs who are looking for work) by the number of people in the labor force (i.e., employed people plus unemployed people). To estimate unemployment rates by occupation, those employed are classified according to their current occupation, and those who are unemployed are classified according to the occupation of their last job held. See <http://www.bls.gov/cps/cps_htgm.pdf> for more detail on how the government calculates the unemployment rate.

[c] Included as a benchmark.

SOURCE: Bureau of Labor Statistics, Current Population Survey, <www.bls.gov/cps/home.htm>.

TABLE D.5 Percent of Workforce That Are U.S. Citizens for NGA-Relevant Occupations

Occupation	Percent U.S. Citizens				
	2005	2006	2007	2008	2009
Computer scientists and systems analysts	89.6	90.0	89.3	89.9	89.9
Computer programmers	89.1	88.6	89.4	89.3	89.3
Computer software engineers	77.9	76.8	79.2	79.0	79.5
Miscellaneous mathematical science occupations, including mathematicians and statisticians	85.8	86.2	85.7	87.1	87.6
Operations research analysts	95.2	95.6	95.9	95.9	95.0
Surveyors, cartographers, and photogrammetrists	97.6	97.7	97.2	98.0	96.0
Aerospace engineers	94.6	94.8	96.5	96.3	96.9
Computer hardware engineers	78.4	87.0	86.5	85.4	88.4
Electrical and electronics engineers	88.9	90.3	90.5	90.6	90.6
Surveying and mapping technicians	98.3	97.2	98.3	97.7	97.1
Astronomers and physicists	91.3	85.2	84.1	83.4	85.9
Atmospheric and space scientists	90.6	99.1	98.9	95.7	95.0
Environmental scientists and geoscientists	95.4	95.5	94.8	95.6	95.1
Physical scientists, all other	78.4	77.0	78.5	76.2	76.3
Urban and regional planners	96.6	98.6	95.0	97.0	96.5
Miscellaneous social scientists, including sociologists	92.9	93.5	93.8	96.3	95.8
Miscellaneous life, physical, and social science technicians	93.0	92.4	92.5	92.5	93.4
Librarians	97.7	97.7	98.1	97.1	97.7
Artists and related workers	92.9	94.0	93.3	93.2	93.2
Statistical assistants	97.1	97.5	97.2	96.2	97.0

SOURCE: Data from the U.S. Census Bureau, American Community Survey, Public Use Microdata Samples (2005-2009). Includes those employed as of the survey reference period. Occupation titles are based on the 2002 Census occupational classification system (and are consistent with the 2000 Standard Occupation Code system).

Appendix E

Biographical Sketches of Committee Members

Keith C. Clarke is a research cartographer and professor in the Geography Department at the University of California, Santa Barbara. He is also the Santa Barbara Director of the National Center for Geographic Information and Analysis. Prior to joining the faculty in 1996, he was a professor at Hunter College and he also spent a year as an advisor to the Office of Research in the U.S. Geological Survey's (USGS's) National Mapping Division. He holds a B.A. from Middlesex Polytechnic (London) and an M.A. and Ph.D. in analytical cartography from the University of Michigan. Dr. Clarke's research focuses on environmental simulation modeling, modeling urban growth, terrain mapping and analysis, and the history of satellite surveillance. He has played numerous leadership roles, including president of the Cartographic and Geographic Information Society and chair of several National Research Council (NRC) committees, including the Committee on Basic and Applied Research Priorities in Geospatial Science for the National Geospatial-Intelligence Agency, the Committee on the New Research Directions for the National Geospatial-Intelligence Agency: A Workshop, and the Mapping Science Committee. Dr. Clarke is a recipient of the John Wesley Powell Award, the USGS's highest award for achievement, and a fellow of the American Congress on Surveying and Mapping.

Luc E. Anselin is Regents' Professor and holds the Walter Isard Chair in the School of Geographical Sciences and Urban Planning at Arizona State University (ASU). He is also the founding director of the School as well as of the GeoDa Center for Geospatial Analysis and Computation at ASU. His Ph.D. in regional science is from Cornell University and he holds a master's degree in econometrics, statistics, and operations research from the Free University of Brussels, where he also obtained an undergraduate degree in economics. Dr. Anselin's research deals with various aspects of spatial data analysis and geographic information science, ranging from exploratory spatial data analysis to geocomputation, spatial statistics, and spatial econometrics. He is a fellow of the Spatial Econometric Society, the Regional Science Association International, and the University Consortium for Geographic Information Science (UCGIS) and was awarded the Walter Isard Award in 2005 and the William Alonso Memorial Prize in 2006. He is a member of the National Academy of Sciences and the American Academy of Arts and Sciences.

Alexandre M. Bayen is an associate professor of systems engineering in the Department of Electrical Engineering and Computer Science, and Civil and Environmental Engineering at the University of California, Berkeley. Prior to joining the faculty, he spent a year as research director of the Autonomous Navigation Laboratory (Ministry of Defense) in France. He holds an engineering degree in applied mathematics from Ecole Polytechnique, France, and an M.S. and Ph.D. in aeronautics and astronautics from Stanford University. His research interests are in mobile Internet applications (location-based services); participatory sensing; inverse modeling and data assimilation; and control, estimation, and optimization of distributed parameter systems. Current sensor-network projects are aimed at measuring water parameters, mapping earthquake shaking, and monitoring traffic. The latter (Mobile Millennium) received the 2008 Best of ITS Award for "Best Innovative Practice" at the ITS World Congress and the TRANNY Award from the California Transportation Foundation. Dr. Bayen received the CAREER award from the National Science Foundation and a Presidential Early Career Award for Scientists and Engineers and was a participant in the 2008 National Academy of Engineering (NAE) Frontiers of Engineering symposium.

Grant C. Black is a teaching professor of economics and director of the Center for Entrepreneurship and Economic Education at the University of Missouri-St. Louis. He was previously an associate professor of economics at Indiana University South Bend, where he also served as director of the Bureau of Business and Economic Research and the Center for Economic Education. He received his Ph.D. in economics from Georgia State University. Dr. Black's research focuses on the economics of science and innovation, including labor markets and training in the sciences, the transfer of knowledge in the economy, the geographic concentration of scientific and innovative activity, and the role of the foreign-born in scientific productivity. He is the author of the book *The Geography of Small Firm Innovation*. Dr. Black served on the Research Team for the NRC Committee for Capitalizing on Science, Technology, and Innovation: An Assessment of the Small Business Innovation Research Program and participated in other studies by the NRC Board on Science, Technology, and Economic Policy. He also participated in activities of the Scientific Workforce Project at the National Bureau of Economic Research and the National Nanotechnology Initiative workshop on societal implications of nanotechnology. He is a fellow of the Institute on the Data Resources of the National Science Foundation.

Barbara P. Buttenfield is a professor of geography at the University of Colorado, Boulder. She also directs the Meridian Lab, a small research facility focusing on visualization and modeling of geographic information and technology. She received her B.A. in geography from Clark University, her M.A. in geography from the University of Kansas, and her Ph.D. in geography from the University of Washington. Dr. Buttenfield's research focuses on map generalization, multiscale geospatial database design, algorithms for web-based data delivery, and visualization of uncertainty in environmental modeling. She has also published on spatial data infrastructures, adoption of geospatial technologies, and digital libraries. While working on her master's degree, she received 12 weeks of training in photogrammetry, photointerpretation, mapping, and charting, and spent a year as a cartographer at the Defense Mapping Agency, a predecessor organization to the National Geospatial-Intelligence Agency. Dr. Buttenfield has served on several NRC committees related to cartography and the mapping sciences, most recently the Committee on Basic and Applied Research Priorities in Geospatial Science for the National Geospatial-Intelligence Agency. She is a past president of the Cartography and Geographic Information Society and a fellow of the American Congress on Surveying and Mapping. In 2001, she was named GIS Educator of the Year by the University Consortium for Geographic Information Science.

Kathleen M. Carley is a professor of computer science at the Institute for Software Research in the School of Computer Science at Carnegie Mellon University. She also directs the university's Center for Computational Analysis of Social and Organizational Systems, which brings together network analysis, computer science, and organization science, and also incorporates a training program for Ph.D. students. She developed and directs the interdisciplinary Ph.D. program in Computation, Organizations and Society. She holds two bachelor's degrees from the Massachusetts Institute of Technology—one in economics and one in political science—and a Ph.D. in sociology from Harvard University. Dr. Carley uses organization theory, dynamic network analysis, social networks, multiagent systems, and computational social science to examine how cognitive, social, technological, and institutional factors affect individual, team, social, and policy outcomes in areas ranging from public health to counterterrorism to cyber security. She also develops tools for analyzing large-scale and geosituated dynamic networks and multiagent simulation systems that are used worldwide. She is a senior member of the Institute of Electrical and

Electronics Engineers (IEEE), received the lifetime achievement award from the Mathematical Sociology section of the American Sociological Association, and the Simmel award for advances in social networks and network science from the International Network for Social Network Analysis. Dr. Carley has participated in several NRC studies, including the Committee on Modeling and Simulation for Defense Transformation and the Panel on Modeling Human Behavior and Command Decision Making: Representations for Military Simulations, and she was a keynote speaker at the 2010 Workshop on New Research Directions for the National Geospatial-Intelligence Agency (NGA).

John R. Jensen is Carolina Distinguished Professor and co-director of the GIS and Remote Sensing Center in the Department of Geography at the University of South Carolina. His research interests are in remote sensing of the environment, digital image processing, and biogeography. He received his B.A. in geography (photogrammetry focus) from California State University, Fullerton; his M.S. in geography (photogrammetry and cartography focus) from Brigham Young University; and his Ph.D. in geography (remote sensing and cartography focus) from the University of California, Los Angeles. He is also a certified photogrammetrist. Dr. Jensen has written four textbooks, including *Remote Sensing of the Environment: An Earth Resource Perspective, Introductory Digital Image Processing: A Remote Sensing Perspective* (now in its third edition), and an electronic book on geospatial processing with interactive frames of instruction and animation. He also serves on education committees and is a current member of the National Center for Geographic Information and Analysis Remote Sensing Core Curriculum Committee and a former chair of the Commission on Education in Remote Sensing and Geographic Information Systems for the International Society for Photogrammetry and Remote Sensing. Dr. Jensen is a former president and current fellow of the American Society for Photogrammetry and Remote Sensing and received that society's Alan Gordon Memorial Award for significant achievements in remote sensing and photographic interpretation.

Richard B. Langley is a professor of geodesy and precision navigation in the Department of Geodesy and Geomatics Engineering at the University of New Brunswick. He received a B.S. in applied physics from the University of Waterloo and a Ph.D. in experimental space science from York University. Dr. Langley has worked extensively on global navigation satellite systems techniques and algorithms for geodetic and high-precision surveying applications and for aircraft navigation and spacecraft systems. He is also interested in the evolving role of geodesy in surveying and mapping education and has given several talks on this topic. Dr. Langley is a co-author of the best-selling *Guide to GPS Positioning* and is a columnist and contributing editor of *GPS World* magazine. He is also active in professional and learned societies associated with geodesy and GPS. He is a past chair of the Canadian National Committee for the International Union of Geodesy and Geophysics and a former member of the European Space Agency GNSS Scientific Advisory Group. He is an elected fellow of the International Association of Geodesy, the Institute of Navigation, and the Royal Institute of Navigation.

Edward M. Mikhail is Professor Emeritus of photogrammetry and the former head of Geomatics Engineering at Purdue University. He holds a B.S. in civil engineering from Cairo University and an M.S. and a Ph.D. in photogrammetry and geodesy from Cornell University. Dr. Mikhail has taught and carried out research in photogrammetry; data adjustment; digital mapping; sensor modeling; and automated methods for feature extraction and analysis, registration, and fusion for more than 46 years. He established Purdue's graduate program in geomatics engineering, and supervised more than 250 master's and some 30 Ph.D. students in photogrammetry and geomatics. He also taught many short courses on various aspects of photogrammetry and mapping to government agencies and private companies. He is familiar with NGA and its needs for photogrammetrists, having spent several sabbaticals as a visiting scientist and training its employees. He currently serves on NGA's technical geopositioning and photogrammetric groups, and supports the Mensuration Services Project. Dr. Mikhail has written many books on photogrammetry, least-squares adjustment, and surveying, and co-edited several manuals, including the *Manual of Photogrammetry* and the *Handbook of Civil Engineering*. He is an honorary member of the American Society for Photogrammetry and Remote Sensing, a distinction held by no more than 25 members at any given time, and received that society's Fairchild Photogrammetric Award for outstanding achievement, as well as the German Alexander von Hombolt Senior Scientist Award. He also received commendations from the U.S. Geological Survey, the Office of Research and Development, and the Imagery Intelligence Directorate of the National Reconnaissance Office.

Shashi Shekhar is the McKnight Distinguished University Professor in the Department of Computer Science at the University of Minnesota. He holds a B. Tech in computer science from the Indian Institute of Technology in Kanpur, India, and an M.S. and Ph.D. in computer science from the University of California, Berkeley. Dr. Shekhar's research interests are in spatial databases and spatial data mining, an interdisciplinary area at the intersection of computer science and geographic information systems. He has co-edited an *Encyclopedia of GIS* and co-authored a textbook on spatial databases. Dr. Shekhar is a fellow of the American Association for the Advancement of Science and IEEE, and received that society's Technical Achievement Award for contributions to spatial database storage methods, data mining, and geographic information systems. He was a member of the NRC Mapping Science Committee, the NRC Committee on Basic and Applied Research Priorities in Geospatial Science for the National Geospatial-Intelligence Agency, and the Board of Directors of the University Consortium for Geographic Information Science.

Michael N. Solem is Director of Educational Affairs at the Association of American Geographers (AAG), where he leads research projects in graduate education, international education, teacher preparation, and workforce development of geographers. He is also directing collaborative projects focusing on trends and issues in geography in K-12 and higher education in the United States and abroad. His publications on these topics appear regularly in the peer-reviewed literature and in conference proceedings. Dr. Solem serves on the International Geographical Union's Commission on Geographical Education and is co-coordinator of the International Network on Learning

and Teaching, which seeks to improve the quality of learning and teaching of geography in higher education internationally. He has twice received the *Journal of Geography in Higher Education*'s biennial award for promoting excellence in teaching and learning for his research on geography faculty development and graduate education. He holds a B.S. in earth sciences from Pennsylvania State University and an M.S. and Ph.D. in geography from Pennsylvania State University and the University of Colorado, Boulder, respectively.

Paula Stephan is a professor of economics in the Andrew Young School of Policy Studies at Georgia State University and a research associate of the National Bureau of Economic Research. She graduated from Grinnell College (Phi Beta Kappa) with a B.A. in economics and earned both her M.A. and Ph.D. in economics from the University of Michigan. Her research interests focus on the careers of scientists and engineers and the process by which knowledge moves across institutional boundaries in the economy. Dr. Stephan has served on a number of NRC committees, including the Committee on Examination of the U.S. Air Force's Science, Technology, Engineering and Mathematics (STEM) Workforce Needs in the Future and Its Strategy to Meet Those Needs; Committee on Dimensions, Causes, and Implications of Recent Trends in the Careers of Life Scientists; Committee on Methods of Forecasting Demand and Supply of Doctoral Scientists and Engineers; and the Committee on Policy Implications of International Graduate Students and Postdoctoral Scholars in the United States. She currently serves on the NRC Board on Higher Education and Workforce, and has been a member of the Scientific Workforce Project at the National Bureau of Economic Research since 2002.

May Yuan is Brandt Professor and Edith Kinney Gaylord Presidential Professor and the director of the Center for Spatial Analysis at the University of Oklahoma. She received a B.S. in geography from National Taiwan University and an M.A. and a Ph.D. in geography from the State University of New York at Buffalo. Dr. Yuan's research interests are in temporal GIS, geographic representation, spatiotemporal information modeling, and applications of geographic information technologies to dynamic systems, such as wildfires and rainstorms. She was the president of the University Consortium for Geographic Information Science, and has served on several committees concerned with geospatial analysis. She is currently a member of the NRC Mapping Science Committee and the Academic Advisory Board for the U.S. Geospatial Intelligence Foundation. She is familiar with geospatial intelligence needs and co-organized a workshop on geographic dynamics sponsored by the intelligence community and produced two books on the subject after the workshop.

Michael J. Zyda is a professor of engineering practice in the Department of Computer Science at the University of Southern California. He also directs the university's GamePipe Laboratory, which engages students in research and development of interactive games. He initiated two cross-disciplinary degree programs—a B.S. in computer science (games) and an M.S. in computer science (game development)—and doubled the incoming undergraduate enrollment of the Computer Science Department. Dr. Zyda is a pioneer in the fields of computer graphics, networked virtual environments, modeling and simulation, and serious games. His research interests include collaboration in entertainment and defense, and he has developed, for example, a game used by the Army for recruiting. He has also served on numerous NRC committees advising the Department of Defense, including the Committee on Modeling and Simulation: Linking Entertainment and Defense and the Committee on Defense Modeling, Simulation and Analysis, and he was a participant in the 2010 Workshop on New Research Directions for the NGA. Dr. Zyda is a National Associate of the National Academies and a member of the Academy of Interactive Arts and Sciences. He received a B.A. in bioengineering from the University of California, San Diego, an M.S. in computer science from the University of Massachusetts, and a Ph.D. in computer science from Washington University, St. Louis.

Appendix F

Acronyms and Abbreviations

3D	three-dimensional
AAG	Association of American Geographers
ACM SIG	Association for Computing Machinery Special Interest Group
AGU	American Geophysical Union
ASPRS	American Society for Photogrammetry and Remote Sensing
CaGIS	Cartography and Geographic Information Society
CASOS	Center for Computational Analysis of Social and Organizational Systems
COS	Computation and Organization Science
CRADA	Cooperative Research and Development Agreement
DHS	Department of Homeland Security
DOD	Department of Defense
ESRI	Environmental Systems Research Institute
GEOINT	geospatial intelligence
GIS	Geographic Information System
GNSS	global navigation satellite systems
GOCE	Gravity field and steady-state Ocean Circulation Explorer
GPS	Global Positioning System
GRSS	Geoscience and Remote Sensing Society
ICPSR	Interuniversity Consortium for Political and Social Research
IEEE	Institute of Electrical and Electronics Engineers
IGERT	Integrative Graduate Education and Research Traineeship
IIF	International Institute of Forecasters
INS	inertial navigation system
INSNA	International Network for Social Network Analysis
ION	Institute of Navigation
ISPRS	International Society for Photogrammetry and Remote Sensing
LTFTT	Long Term Full Time Training
MOOC	Massive Open Online Course
NASA	National Aeronautics and Space Administration
NCGIA	National Center for Geographic Information and Analysis
NGA	National Geospatial-Intelligence Agency
NIMA	National Imagery and Mapping Agency
NRC	National Research Council
NSF	National Science Foundation
NWS	National Weather Service
SMART	Science, Mathematics, and Research for Transformation
SPIE	Society of Photographic Instrumentation Engineers
UARC	University Affiliated Research Center

UCGIS   University Consortium for Geographic   USC   University of Southern California
         Information Science   USGIF   U.S. Geospatial Intelligence Foundation
URISA   Urban and Regional Information Systems
         Society   WGS 84   World Geodetic System 1984